T0073197

Handbook of Boron Nanostructures

Handbook of Boron Nanostructures

edited by
Sumit Saxena

PAN STANFORD PUBLISHING

Published by

Pan Stanford Publishing Pte. Ltd.
Penthouse Level, Suntec Tower 3
8 Temasek Boulevard
Singapore 038988

Email: editorial@panstanford.com
Web: www.panstanford.com

British Library Cataloguing-in-Publication Data
A catalogue record for this book is available from the British Library.

Handbook of Boron Nanostructures
Copyright © 2016 by Pan Stanford Publishing Pte. Ltd.

ISBN   978-981-4613-94-1 (Hardcover)
ISBN   978-981-4613-95-8 (eBook)

Printed in the USA

Contents

Preface   ix

1. Introduction to Boron Nanostructures   **1**
Sumit Saxena

   1.1   Introduction   1
   1.2   Boron Nanoclusters   3
   1.3   Boron Nanosheets   4
   1.4   Boron Nanoribbons   7
   1.5   Boron Nanotubes   8
   1.6   Boron Nanowires   10

2. Boron Nanoclusters   **13**
Saurabh Awasthi and Sumit Saxena

   2.1   Introduction   13
   2.2   Development of Boron Nanostructures   16
   2.3   Conclusion and Outlook   17

3. Two-Dimensional Boron Sheets   **19**
Sohrab Ismail-Beigi

   3.1   Introduction   19
   3.2   Atomically Thin Boron Sheets   21
   3.3   Mixed-Phase Triangular-Hexagonal
            Boron Sheets   26
   3.4   Stability of Mixed-Phase Boron Sheets   30
   3.5   Buckling of Boron Sheets   37
   3.6   Double-Layered Boron Sheets   41
   3.7   Interlayer Bonding in Double-Layered
            Boron Sheets   41
   3.8   Outlook   43

4. Ab initio Study of Single-Walled Boron Nanotubes   **49**
Kah Chun Lau and Ravindra Pandey

   4.1   Introduction   49

4.2	The Basic Construction Model of Single-Walled Boron Nanotubes	52
	4.2.1 The Single-Layer Two-Dimensional Boron Sheets	53
4.3	Basic Properties of an SWBNT Derived from an α-Boron Sheet	58
	4.3.1 Basic Geometry Construction	58
	4.3.2 Energetic Stability	59
	4.3.3 Electronic Properties	62
	4.3.4 Elastic Properties	67
4.4	Conclusion and Outlook	70

5. Boron Nanowires: Synthesis and Properties **77**
Shobha Shukla

5.1	Introduction	77
5.2	Synthesis of Boron Nanowires	79
	5.2.1 Chemical Vapor Deposition	81
	5.2.2 Magnetic Sputtering	85
	5.2.3 Laser Ablation	88
5.3	Properties of Boron Nanowires	92
	5.3.1 Electrical Transport Properties	92
	5.3.2 Field Emission Properties	94
5.4	Outlook	96

6. Applications of Boron Nanostructures in Medicine **101**
Komal Sethia and Indrajit Roy

6.1	Introduction	101
6.2	Boron Neutron Capture Therapy	102
6.3	Various Boron-Containing Nanoparticles in Medicine	104
	6.3.1 Pure Boron Nanoparticles	104
	6.3.2 Boron-Containing Nanoparticles	105
	6.3.2.1 Boron carbides	105
	6.3.2.2 Boron nitride nanotubes	107
	6.3.2.3 Nanoclusters of boron with iron	112
	6.3.2.4 Boron–organic hybrid nanoparticles	112

	6.3.2.5 Boron nanoparticles associated with polymeric micelles	113
6.4	Other Applications of Boron Nanoparticles	115
	6.4.1 Boron Nanoparticles in Tissue Engineering	115
	6.4.2 Boron Nanoparticles in Biosensing	115
6.5	Conclusion	116
Index		123

Preface

The success story of carbon-based materials such as fullerenes, nanotubes, and graphene has motivated the exploration of similar material systems. Several material systems are being explored with the expectation of discovering novel phenomena leading to the development of smart devices. Being almost similar to carbon, boron, thanks to its own peculiar properties, forms one of the most obvious choices in this pursuit. Several attempts are being made to harness the potential of this emerging material system, and efforts are under way to understand boron-based materials systems both theoretically and computationally.

The purpose of this book is to bring together the knowledge base generated by several researchers working in this exciting area. The book is intended for young researchers and provides a succulent summary of important theoretical and experimental developments in the area of nanoboron.

Sumit Saxena
February 2016

Chapter 1

Introduction to Boron Nanostructures

Sumit Saxena
Department of Metallurgical Engineering and Materials Science,
Indian Institute of Technology Bombay, Mumbai 400076, India
sumit.saxena@iitb.ac.in

1.1 Introduction

Boron is a rare element both in the solar system as well as in the earth's crust. It is produced by cosmic ray spallation and since the geochemical processes have greatly concentrated boron, it is not considered as a rare element from the perspectives of commercial availability. One of the earliest accounts of the use of boron has been found in the Chinese civilization in the form of borax glazes since AD 300. Boron was first partially isolated in 1808 by French chemist Joseph L. Gay-Lussac and L. J. Thenard [1] by reducing boric acid at high temperature and independently by Sir Humphry Davy [2] using electrolysis.

Handbook of Boron Nanostructures
Edited by Sumit Saxena
Copyright © 2016 Pan Stanford Publishing Pte. Ltd.
ISBN 978-981-4613-94-1 (Hardcover), 978-981-4613-95-8 (eBook)
www.panstanford.com

Boron occurs in two naturally occurring isotopes, B_{10} and B_{11}. Although the B_{11} isotope forms the major chunk of the naturally existing isotopes, the existence of the B_{10} isotope is of considerable significance due to its large cross section for thermal neutron capture. The boron atom has five electrons and its electronic configuration in the ground state can be represented as below. The singlet state can be observed in borylene (:B-H). In terms of electron orbitals, this six-electron molecule has a core shell (1σ), a bonding pair (2σ), and a lone pair of electrons (3σ) on the boron atoms [3, 4].

Boron in most stable compounds exhibits an oxidation state of +3, which is achieved by promoting an electron from the $2s$ to the $2p$ orbital.

This causes strong electron acceptor properties of boron compounds and due to the very high ionization potentials, boron (III) does not show the chemistry of B^{3+} ions but that of covalent boron (III) compounds. The tetra-coordinated boron compounds, depending on the ligands, are found to be observed in negative, neutral, and positive charge states. The conventional localized bond between two atoms uses an atomic orbital from each atom and is filled by two electrons. This does not hold true for electron-deficient compounds such as that of boron [5]. The bonding in these compounds has been described by Eberhardt, Crawford, and Lipscomb by using a three-center localized molecular orbital employing three orbitals but a single electron pair [6]. The two-electron, three-centered bonding in boron has been able to provide insights into the structural and electronic properties of boron nanostructures.

This chapter introduces different boron nanostructures such as nanoclusters, nanosheets, nanotubes, and nanowires that have been either experimentally realized and/or theoretically investigated and have been discussed in details in subsequent chapters.

1.2 Boron Nanoclusters

The study of nanoboron chemistry has been catalyzed since the discovery of the B_{13+} nanocluster by Hanley and Anderson [7]. Since then a large number of reports, both experimental and theoretical, have surfaced, investigating their structure [8, 9], stability [10], and interactions [11]. Theoretical investigations based on ab initio quantum chemical techniques by Boustani [12] have predicted the ground-state structural configurations of small boron nanoclusters to be planar, as illustrated in Fig. 1.1. This elemental set comprises of boron clusters B_n, where n = 2–8. The symmetries of different elemental clusters are C_{2v} for B_3, B_5, and B_7, C_{5v} for B_6, D_{7h} for B_8, D_{2h} for B_4, and planar B_6 nanoboron clusters.

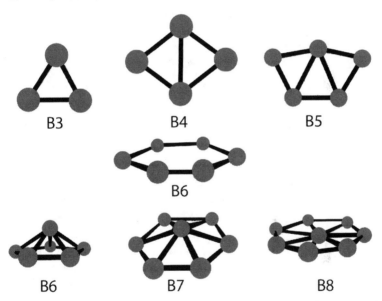

Figure 1.1 Ground-state structures along with structural symmetry of neutral boron nanoclusters obtained using the 3-21G basis set [12].

This set of cluster forms the structural basis for larger nanoclusters. The set of quasiplanar and convex clusters is mainly composed of units of hexagonal pyramids which belong to this elemental set of clusters. The B_6 clusters are found to exist in planar hexagonal and pentagonal pyramid conformations and form the most stable elements of this set. Larger nanoclusters of boron acquire quasiplanar and very compact 3D structures, as shown in Fig. 1.2 below. Quasiplanar boron nanoclusters can be understood to be constructed from hexagonal pyramidal subunits by dovetailing each other and forming axial bonds; however, when the out-of-plane apices of these subunits either lie above the plane of peripheral atoms, such nanoclusters are known as convex clusters. Quasiplanar boron nanostructures are obtained when the apices of subunits lie alternatively above and below the plane containing peripheral atoms. The structures of 3D nanoclusters are similar to well-known α- and β-rhombohedral boron crystals.

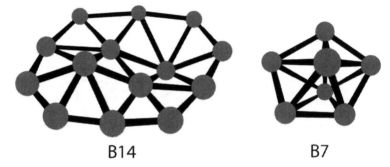

B14   B7

Figure 1.2   Representative quasiplanar and 3D nanoclusters of boron.

The 3D nanoclusters can be classified into compact 3D and open 3D nanoclusters. The very compact 3D clusters ($n <= 9$) consists of trigonal, square, pentagonal, hexagonal, and heptagonal bipyramids. The open 3D structures ($n = 10\text{--}14$) have low point group symmetries. Ab initio calculations at the Hartree–Fock self-consistent field (HF-SCF) level have suggested that the energies of 3D nanoclusters are higher than those of convex or quasiplanar clusters. The insight into the structure of boron nanoclusters is expected to pin down the atomic structure of higher-dimensional boron nanostructures.

1.3   Boron Nanosheets

With the successful interpretation of the properties of carbon nanostructures using graphene, the 2D nanosheets are now understood to hold the key for understanding the properties of nanomaterials in general. The experimental discovery of boron nanotubes has motivated the investigation of the structural properties of monolayer boron sheets. Growth of a monoatomic layer of boron nanosheets on a silicon(100) substrate havs been reported by Weir et al. [13]. It is observed that 0.5 monolayer coverage forms an ordered 2 × 1 structure and can be distinguished using a low-energy electron diffraction (LEED) pattern. The LEED pattern for pristine silicon shows (1/2, 0) as the most intense half-order, whereas (1, 1/2) forms the most intense half-order spot in the diffraction pattern from a boron-covered surface [14] (Fig. 1.3).

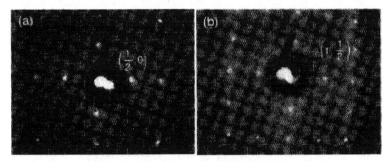

Figure 1.3   Low-energy electron diffraction patterns for (a) pristine silicon (100)-2 × 1 and (b) silicon (100)-2 × 1 with a 0.5 monolayer of boron [14].

Investigation of the local structure using scanning tunneling microscopy along with tunneling spectroscopy suggests that the principal structural subunit is an ordered arrangement of four boron atoms at substitutional sites in the first bulk silicon-like layer, which is then capped with an ordered arrangement of silicon dimmers and dimmer vacancies [15] (Fig. 1.4). These ordered surface reconstructions are stable under vacuum and can be preserved as metastable states at the solid–solid interfaces [16]. A lot of effort has been made to synthesize and study the structural properties of monolayer boron sheets; however, no experimental verification of stabilized monolayer boron sheets under normal conditions has been reported so far. Different atomic models of monolayer boron sheets such as the idealized and buckled triangular {1212} [17, 18], reconstructed {1221} [19], and hexagonal sheets [20] have been proposed using computational methods (Fig. 1.5).

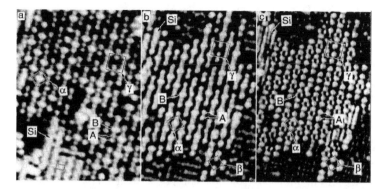

Figure 1.4 High-resolution images of boron induces reconstructions on a Si(001) surface α, β, and γ denote the α–c (4 × 4), β–c (4 × 4), and γ (4 × 4) unit cells of observed reconstructions. A and B denote the structural subunits of reconstructions. (a) Reconstructions and region of clean Si(001) (2 × 1) outlined. (b) High-resolution image of occupied surface electronic states. (c) Unoccupied electronic states [15].

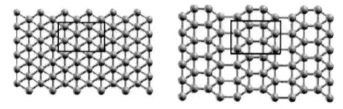

Figure 1.5 Atomic structure of proposed idealized triangular {1212} and reconstructed {1221} monolayer boron sheet [17, 18].

One of these models, particularly the hexagonal monolayer boron sheet, has been inspired by the atomic model of graphene sheets. First-principles calculations have recently indicated the α-boron sheets to be the most stable ground structure of monolayer boron sheets (Fig. 1.6).

These sheets form precursors to B_{80} clusters and conduct only by means of out-of-plane p_z orbitals [21]. The holes in the α-boron sheets are understood to serve as scavengers of extra electrons from the adjacent filled hexagons [22].

The atomic structures of these 2D nanostructures have been discussed in detail in Chapter 2, and a comprehensive understanding of the structure will enable further understanding of quasi-2D and other higher-dimensional boron nanostructures.

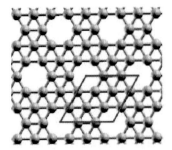

Figure 1.6 Atomic structure of a proposed α-boron sheet using density functional calculations [21].

1.4 Boron Nanoribbons

Quasi-2D nanostructures hold the key to the realization of nanoelectrical devices, specifically nanojunctions. These nanoribbons can be produced using either top-down or bottom approaches. Carbon nanoribbons are the most explored quasi-2D nanostructures and have been realized by controlled unzipping single-walled carbon nanotubes [23], lithography [24], and wet-chemical methods [25]. One of the first reports on synthesis of single-crystalline α-boron nanoribbons was published by Xu et al. [26] by catalyst-free pyrolysis of diborane at 630°C–750°C at low pressures. Nanoribbons synthesized using this method have recently been investigated for thermal transport properties.

The thermal conductivity of bundled boron nanoribbons is higher than that of a single nanoribbon, and more importantly their thermal transport properties can be reversible modified by wetting the van der Waals interface between nanoribbons and various solutions [27]. The properties of boron nanowires have been reported using first-principles density functional theory. These nanoribbons were derived by cutting α-boron sheets along different directions, producing linear and armchair-edged nanribbons. A comparative study of the electron localization function shows that the delocalization of π-electron orbitals occurs more in nanoribbons derived from •-boron sheets (Fig. 1.7) than in those derived from other precursors [28]. These preliminary reports on boron nanoribbons using ab initio methods and experimental investigations suggest that nanodevices capable of thermal management can be fabricated using these nanostructures.

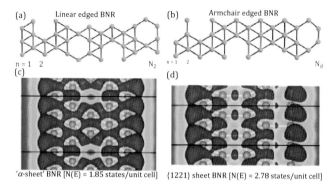

'α-sheet' BNR [N(E) = 1.85 states/unit cell]   {1221} sheet BNR [N(E) = 2.78 states/unit cell]

Figure 1.7   Atomic structure of nanoribbons derived from an α-boron sheet. The ELF plot for nanoribbons derived from (a) a stable α-boron and (b) a reconstructed {1221} sheet [28].

1.5   Boron Nanotubes

Boron nanotubes are a new class of materials that have been theoretically predicted to exist in stable form; however, the exact atomic structure and hence their electronic properties are a subject of discussion. One of the first accounts of successful synthesis of boron nanotubes was reported by Ciuparu et al. using a catalyst template–assisted growth technique [23]. These samples were investigated using Raman spectroscopy and showed phonon modes to suggest their tubular structures (Fig. 1.8).

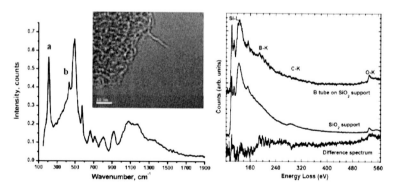

Figure 1.8   (a) Raman spectra of a boron nanotube. The inset shows the micrograph of the boron nanotube. (b) Electron energy loss spectrum of a boron nanotube on a SiO_2 substrate [23].

The presence of vibration modes below 500 cm^{-1} in the Raman spectra have been understood due to the presence of radial breathing modes in boron nanotube samples. The electron energy loss spectra suggest the presence of B K-edge, indicating its origin from the boron tubular nanostructures.

There have been a lot of speculations on the atomic structure of boron nanotubes due to the unavailability of any clear-cut characterization reports. A large number of first-principles investigations using density functional theory have been reported time and again in order to solve this mystery. Amongst the earliest investigations, idealized triangular [18, 24], hexagonal, and reconstructed {1221} boron sheets were considered as precursors to boron nanotubes. With the recent predictions of α-boron sheets [21] as being the most stable monolayer sheets, the properties of boron nanotubes formed by rolling these sheets have been investigated. Ab initio investigations suggest that the radial breathing modes of nanotubes formed by rolling an α-boron sheet depend strongly upon the chirality and hence the diameter of the nanotubes, as summarized in Table 1.1.

Table 1.1 Calculated radial breathing mode frequencies of different boron nanotubes using ab initio methods [25]

Nanotube	Diameter(Å)	f_{RBM} (cm^{-1})
(5, 5)	8.13	238.9
(9, 0)	8.63	224.8
(6, 6)	9.93	195.9
(12, 0)	11.31	170.6
(7, 7)	11.37	173.7
(8, 8)	13.13	150.2
(18, 0)	16.50	119.2

The calculated frequencies are in close agreement with the experimentally observed frequency of 210 cm^{-1}, as reported by Ciuparu et al. [23]. The unavailability of experimental data has created a bottleneck in verification of theoretical prediction.

Further insight into the atomic structure based on density functional theory calculations has been developed in Chapter 3. This will enable us in understanding and exploiting the properties of these exotic nanostructures for device applications.

1.6   Boron Nanowires

Nanowires are the most investigated boron nanostructures experimentally. They have been synthesized both in crystalline and amorphous forms using different growth techniques such as chemical vapor deposition [26, 27], magnetic sputtering [28] techniques, and laser ablation [29, 30] of targets using pulsed lasers (Fig. 1.9). Experimental results clearly indicate that the presence of catalytic elements is one of the most crucial conditions for the growth of crystalline boron nanowires.

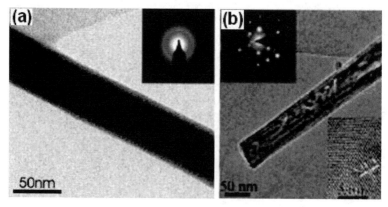

Figure 1.9   HRTEM images of (a) amorphous boron nanowires synthesized using the high-temperature laser ablation method. The inset shows amorphous halo rings in the SAED pattern [30] and (b) crystalline boron nanowires synthesized using the chemical vapor deposition technique. The bright spots in the inset on the top show a SAED pattern, indicating the single-crystalline nature of nanowires, while the inset below shows a lattice spacing of 0.511 nm, which matches with the {121} facet of the rhomb-centered hexagonal boron structure [27].

Crystalline boron nanowires are understood to be semiconducting near room temperatures. The conductivities of these nanowires is very robust and does not changes even on application of large strain, indicating that these nanostructures are highly resilient, even under highly strained conditions [31]. These nanostructures also exhibit excellent field emission properties.

Further details describing the synthesis of boron nanowires and models to understand their growth kinetics and properties have been discussed in Chapter 5. This is expected to provide further insight into using these nanostructures, particularly in field emission devices and devices capable of operating under extreme conditions.

References

1. Gay Lussac, J. L., and Thenard, L.J., *Annales de chimie*, **68**, 169 (1808).
2. Davy H., *Philosophical Transactions of the Royal Society of London*, **99**, 33 (1809).
3. Blint R. J., and Goddard III W. A., *Chemical Physics*, **3**, 297 (1974).
4. Pople J. A., and Schleyer P. v-R., *Chemical Physics Letters*, **129**, 279 (1986).
5. Dickerson R. E., and Lipscomb W. N., *Journal of Chemical Physics*, **27**, 212 (1957).
6. Eberhardt, C., and Lipscomb J., *Journal of Chemical Physics*, **22**, 989 (1954).
7. Hanley L., and Anderson S.L. *Journal of Physical Chemistry*, **91**, 5161 (1987).
8. Hanley L., and Anderson S.L. *Journal of Physical Chemistry*, **92**, 5803 (1988).
9. Ray A. K., Howard I. A., and Kanal K. M., *Physical Review B*, **45**, 14247 (1992).
10. Kawai R., and Weare J. H., *Chemical Physics Letters*, **191**, 311, (1992).
11. Sowa M. B., Snoalnoff A. L. Lapicki A., and Anderson, *Journal of Chemical Physics*, **106**, 9511 (1997).
12. Boustani I., *Physical Review B*, **55**, 16426 (1997).
13. Weir B. E., Headrick R. L., Shen Q., Feldman L. C., Hybertsen M. S., Needels M., Schluter M., and Hart T. R., *Physical Review B*, **46**, 12861 (1992).
14. Headrick R. L., Weir B. E., Levi A. F. J., Eaglesham D. J., and Feldman L.C., *Applied Physics Letters*, **57**, 2779 (1990).
15. Wang Y., Hamers R. J., and Kaxiras E., *Physical Review Letters*, **74**, 403, (1995).
16. Headrick R. L., Weir B. E., Levi A. F. J., Freer B., Bevk J., and Feldman L.C., *The Journal of Vacuum Science and Technology*, **9**, 2269 (1991).

17. Kunstmann J., Quandt A., *Physical Review B*, **74**, 035413 (2006).

18. Cabria I., Alonso J. A., Lopez M. J., *Physica Status Solidi (a)*, **203**, 1105 (2006).

19. Lau K. C., Pati R., Pandey R., and Pineda A. C., *Chemical Physics Letters*, **418**, 549 (2006).

20. Lau K. C., and Pandey R., *The Journal of Physical Chemistry C*, **111**, 2906 (2007).

21. Tang H., and Ismail-Beigi S., *Physical Review Letters*, **99**, 115501 (2007).

22. Galeev T. R., Chen Q., Guo J.-C., Bai H., Miao C.-Q., Lu H.-G., Sergeeva A. P., Li S.-D., and Boldyrev A. I., *Physical Chemistry Chemical Physics*, **13**, 11575 (2011).

23. Wei D., Xie L., Lee K. K., Hu Z., Tan S., Chen W., Sow C. H., Chen K., Liu Y., and Wee A. T. S., *Nature Communications*, **4**, 1374 (2012).

24. Han M. Y., Ozyilmaz B., Zhang Y., and Kim P., *Physical Review Letters*, **98**, 206805 (2008).

25. Cai J. M., Ruffieux P., Jaafar R., Bieri M., Braun T., Blankenburg S., Muoth M., Seitsonen A. P., Saleh M., Feng X. L., Mullen K., and Fasel R., *Nature*, **466**, 470 (2010).

26. Xu T. T., Zheng J.-G., Wu N., Nicholls A. W., Roth J. R., Dikin D. A., and Ruoff R. S., *Nano Letters*, **4**, 963 (2004).

27. Yang J., Yang Y., Waltermire S. W., Wu X., Zhang H., Gutu T., Jiang Y., Chen Y., Zin A. A., Prasher R., Xu T. T., and Li D., *Nature Nanotechnology*, **7**, 91 (2012).

28. Saxena S., and Tyson T. A., *Physical Review Letters*, **104**, 245502 (2010).

29. Ciuparu D. Klie R. F., Zhu Y., and Pfefferle L., *The Journal of Chemical Physics B*, **108**, 3967 (2004).

30. Evans M. H., Joannopoulos J. D., and Pantelides S. T., *Physical Review B*, **72**, 045434 (2005).

31. Singh A. K., Sadrzadeh A., and Yakobson B. I., *Nano Letters*, **08**, 1314 (2008).

Chapter 2

Boron Nanoclusters

Saurabh Awasthi and Sumit Saxena
Department of Metallurgical Engineering and Materials Science,
Indian Institute of Technology Bombay, Mumbai 400076, India
sumit.saxena@iitb.ac.in

2.1 Introduction

Cluster science might be new but it is still an emerging field with unprecedented properties and plausible applications. A cluster of atoms may contain from a few atoms to thousands of atoms in it, with varied structural forms such as tubes, spheres, cones, etc. Cluster science has gained tremendous interest as the properties exhibited by clusters are often enigmatic and different from their bulk counterparts. Elemental boron is synonymous with polymorphs and is as interesting as its neighbor carbon due to its ability to form structures such as nanotubes, fullerenes, planar sheets, etc. Amorphous boron was synthesized for the first time in 1808 via reduction of boron oxide with potassium. The study of chemical bonding between atoms of boron is of rudimental importance, leading to an understanding of enigmatic boron polymorph

Handbook of Boron Nanostructures
Edited by Sumit Saxena
Copyright © 2016 Pan Stanford Publishing Pte. Ltd.
ISBN 978-981-4613-94-1 (Hardcover), 978-981-4613-95-8 (eBook)
www.panstanford.com

properties such as superconductivity, superhardness, and low compressibility. Two polymorphs of boron, namely, α-rhombohedral boron [1] (Fig. 2.1) and β-rhombohedral boron, can be obtained at normal pressure. Recently synthesis of a new high-temperature and high-pressure polymorph, orthorhombic γ-B_{28}, has been reported by Mondal et al. [2]. This is reported to be stable at high-pressure and high-temperature conditions but the same cannot be said about α- and β-porlymorphs of boron.

B_{12}

Figure 2.1 Crystal lattice of α-rhombohedral boron [1].

The fundamental building block of all boron polymorphs is a 12-atom cluster (B_{12}) having icosahedral geometry. In boron, the covalent bonding between B_{12} clusters is either direct or via additional atoms. Theoretical calculations [2, 3] exhibit that B_{12} clusters have 13 internal and 12 external bonding orbitals. According to the Wade–Jemmis rule [4–6], in a B_{12} cluster 26 out of 36 valence electrons are accommodated in 13 internal bonding orbitals to form a cluster. This leaves 10 valence electrons and 12 external bonding orbitals, creating an electron deficiency. This discrepancy is solved by accommodation of valence electrons in two-electron, three-center (2e3c) bonds and one-electron, two-center (1e2c) bonds with normal two-electron, two-center (2e2c) bonds. So, chemical bonding in these structures is characterized by polycenter bonds on the B_{12} closo-cluster and 2e2c and 2e3c bonds between the clusters. The γ-B_{12} structure and its geometrical arrangement provide significant insight into the boron structure.

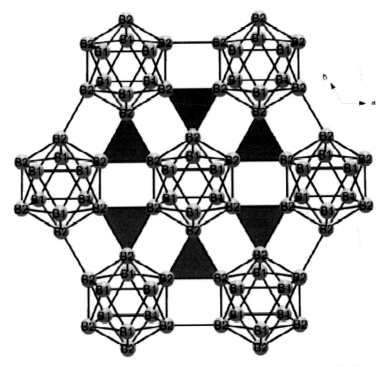

Figure 2.2   View of a B_{12} along the c axis. The atoms B_1 and B_2 are crystallographically independent with the 2e3c bonds of the central B_{12} cluster highlighted in red color [6].

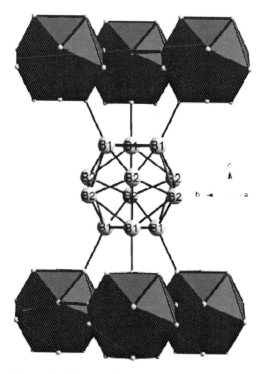

Figure 2.3   View of the crystal structure of •-B_{12}, showing the six intercluster 2e2c bonds for the central B_{12} cluster [1].

The γ-B_{12} structure can be understood as a distorted cubic closed packing of icosahedral B_{12} clusters as spheres. In this each B_{12} cluster is connected to six neighboring B_{12} units by three-center (3c) bonds. The geometrical arrangement is such that these are perpendicular to the threefold symmetry axis within the same close-packed layer, as shown in Fig. 2.2. Then each cluster is furthermore connected to three B_{12} clusters in the layer above and three clusters in the layer below through two-center (2c) bonds, as shown in Fig. 2.3.

2.2   Development of Boron Nanostructures

Stocks pioneering research prepared compounds B_nH_{n+4} (where $n = 2, 5, 6$, or 10) and B_nH_{n+6} ($n = 4$ or 5) and were labeled "electron deficient" as they contained too few electrons in order to form pairs in the molecular structure. Containing $(2n+4)$ or $(2n+6)$ atoms, respectively, these boranes have only $(2n+2)$ or $(2n+3)$ valence shell electron pairs instead of $(2n+3)$ or $(2n+5)$ required for 2c links. R. P. Bell and H. C. Longuet-Higgins provided a structural breakthrough in borane structure development by deducing the structure of diborane, B_2H_6, from its vibrational spectrum. Subsequently, these developments lead to the inception of the 2e3c bond concept and its importance in the borane molecular bonding.

The low-temperature X-ray crystallographic studies of B_nH_{n+m} by W. N. Lipscomb and analysis of its structure put borane chemistry on a sound basis. Lipscomb's studies reflected that n boron atoms and m (endo) hydrogen atoms stay on the inner near-spherical surface, whilst the remaining (exo) hydrogen atoms are accommodated at the outer surface, attached to boron atoms by 2e2c B–H bonds. The inner B–H bonding (endo) evolves $(2n+m)$ electrons, which he characterized into four types of electron bond pairs, namely, sBHB and tBBB 2e3c bonds and yBB and xBH 2e2c bonds.

2.3 Conclusion and Outlook

In short, there is an intriguing and potentially significant difference between carbon and boron due to their fundamental difference in valence electrons of individual atomic orbitals. From its basic electron-deficient feature, it is reasonable to assume that a 2D boron sheet is generally a "frustrated" system which does not have enough electrons to fill all electronic orbitals in a chemical bonding that is based on pure sp^2 hybridization, and therefore, it is highly probable it consequently does not exhibit some clear preference for a simple structural motif that is similar to graphene or other allotropes of carbon. Hence, from an energetics perspective, there is no driving force motivating boron to form a well-defined structure, and this might explain the difficulty in the synthesis of single-walled boron nanotubes (SWBNTs) or single-layer boron sheets. From the fact that no elemental boron but only compounds containing boron can be found on earth, together with the nature of their electron-deficient chemical bonds, might indicate that these boron structures are chemically reactive and will therefore not easily be found under ambient growth conditions. Thus, future studies on the catalyst-promoted growth environment seem to be necessary.

References

1. Quandt A., and Boustani I., *ChemPhysChem*, **6**, 2001 (2005).

2. Mondal S., van Smaalen S., Parakhonskiy G., Prathapa S. J., Noohinejad L., Bykova E., and Dubrovinskaia N., *Physical Review B*, **88**, 024118 (2013).

3. Longuet-Higgins H. C., and de V. Roberts M., *Proceedings of the Royal Society of London A*, **230**, 110 (1955).

4. Lipscomb W. N., and Britton D., *Journal of Chemical Physics*, **33**, 275 (1960).

5. Wade K., *Journal of the Chemical Society, Chemical Communications*, 792 (1971).

6. Wade K., *Advances in Inorganic Chemistry and Radiochemistry*, **18**, 1 (1976).

Chapter 3

Two-Dimensional Boron Sheets

Sohrab Ismail-Beigi
Department of Applied Physics, Yale University, New Haven, CT 06520, USA
sohrab.ismail-beigi@yale.edu

3.1 Introduction

This chapter's focus is on the relationship between the atomic-scale structure, bonding motifs, energetic stability, and electronic structure of extended 2D boron sheets whose thickness is at the atomic scale. How does a particular bonding pattern change these properties and what range of possible properties can be obtained for 2D boron at the nanoscale? And what are the most stable bonding patterns and why?

There are a number of reasons why one would care to study 2D boron sheets, as shown in some of the other chapters in this book. For example, much like for carbon nanotubes or fullerenes where many of their properties can be understood by viewing them as curved graphene sheets, stable 2D boron sheets are reasonable starting materials for understanding the structure and properties

Handbook of Boron Nanostructures
Edited by Sumit Saxena
Copyright © 2016 Pan Stanford Publishing Pte. Ltd.
ISBN 978-981-4613-94-1 (Hardcover), 978-981-4613-95-8 (eBook)
www.panstanford.com

of analogous boron nanotubes or hollow fullerene-like clusters. However, graphene, namely, 2D carbon, itself has proven to be an enormously interesting material in its own right [1] (see also http://www.nobelprize.org/nobel_prizes/physics/laureates/2010/), so studying a 2D material can provide a scientific impact that stands on its own. Viewed this way, one can ask pragmatically, How do 2D boron sheets differ from graphene and what new properties do they bring to the table?

One way to approach the structure and properties of boron sheets is to begin with small boron clusters (i.e., homopolar molecules) and then to steadily increase their size to approach the extended limit. There is a large body of literature on this subject, at first purely theoretical (some of the original work is represented in Refs. [2–7] and then with significant experimental input [8–12]. Generally, for small clusters flat planar motifs are found that may then turn into nanotubular geometries for larger clusters. More ambitious studies identified possibly stable fullerenes such as B_{80} with hollow interiors [13–16]. However, we would like to warn that once the boron clusters get slightly larger (somewhere above 40 atoms), the hollow structures are no longer the ground state by a significant energy cost and instead the stable structures contain an icosahedral core surrounded by a disordered and low-symmetry outer shell [17–19]. Therefore, this theoretical approach is not easy to generalize to larger clusters in a straightforward manner; experimentally, one would guess that trapping the cluster in the higher-energy fullerene or nanotubular geometries is difficult.

In this chapter, we begin directly with the extended limit of large, theoretically infinite and periodic, 2D boron sheets. Thus there are no boundaries or edges to worry about, and one can focus directly on the bonding and properties of the interior of the sheet itself. Of course, a 2D extended sheet will not be the ground state for boron as it prefers 3D bulk crystal phases. So we must assume the 2D structure is stabilized in some manner, for example, via a suitable substrate or by geometrical confinement, and then proceed to examine the properties of the sheets.

Readers will notice that this chapter describes results that are theoretical, as presently there is no known experimental realization of extended 2D sheets of boron that are one or two atoms thick. We will return to this issue in the Outlook section at the end of this chapter. The theoretical methods used are all based on ab initio or first-principles density functional theory (DFT). In brief, DFT is an approach for computing the total energy and electronic distribution of any configuration of atoms in its electronic ground state; ab initio means that there are no free or adjustable parameters, since the computations only involve fundamental constants and the atomic numbers of the constituent atoms. For these reasons, the results of DFT calculations are expected to be *reliable* in that they should have relatively uniform errors across a wide range of atomic bonding geometries that may or may not be bulk-like, making DFT highly valuable when studying novel or nanoscale materials. The question of whether the approximations one must make in practice to perform DFT calculations yield results that are *accurate* enough to be useful in materials predictions is obviously a critical but separate question. For materials such as boron that are not normally expected to have strong electronic correlations, it is widely believed that standard approximations such as the local density or generalized gradient approximations are sufficiently accurate, and for known bulk or molecular phases of elemental boron and boron compounds, the predictions from DFT are certainly in good to excellent agreement with experiments. We will not be describing DFT calculations or the associated toolboxes in this chapter since standard reviews of these well-developed topics can be easily found elsewhere.

3.2   Atomically Thin Boron Sheets

We begin with extended 2D sheets of pure boron that are a single atom thick, that is, boron analogues of graphene. Mathematically, we will be considering the infinite limit of periodic 2D boron sheets so that we can concentrate on the bonding in the interior regions. A fruitful way to begin is to start with the honeycomb arrangement of graphene (see Fig. 3.1 for its electronic structure and Fig. 3.2 for the atomic geometry). Since the main difference between boron and carbon atoms as neighbors on the periodic table is their electron count, the electronic band structure of honeycomb or hexagonal boron sheets is very similar to that of graphene. Looking at Fig. 3.1, we note (i) low-energy in-plane or σ bonding states separated by an energy gap from the high-energy in-plane σ^* antibonding states,

(ii) out-of-plane bonding π and antibonding π^* states separated by a precise zero in the densities of states (DOSs) (the famous Dirac point for graphene which is a special nongenetic feature of the hexagonal lattice), and (iii) the Fermi-level cuts through the low-energy bonding σ and π states, making them both partially unoccupied. Features (i) and (ii) are inherent to the structure of the DOS of the 2D hexagonal lattice and are shared with graphene, while feature (iii) is due to the lower electron count of B compared to C. In short, B does not have enough electrons to occupy all the low-energy and stabilizing bonding states of this atomic arrangement. And this in turn leads to the fact that boron in the hexagonal structure is mechanically unstable [6, 7, 20].

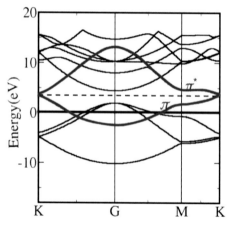

Figure 3.1 Band structure of hexagonal 2D boron from density functional theory within the local density approximation along high-symmetry directions. Solid black lines are in-plane or σ states, while the blue curves are out-of-plane or π states. The bonding π and antibonding π^* bands meet at single Dirac points (K point) in the Brillouin zone; the σ and σ^* bands are separated by a band gap at the gamma point (G). The horizontal dashed curve is the Fermi level for graphene (i.e., if the atoms were carbon), whereas the solid horizontal line at zero energy is the actual Fermi level for boron.

For what follows, it is helpful to review how these bands are formed from localized orbitals. For the hexagonal lattice with threefold symmetry about each site, the use of sp^2 hybrids forms a natural description. As shown in Fig. 3.2, one can consistently orient the three sp^2 hybrids on each atom to point directly at its nearest neighbor.

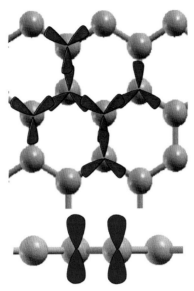

Figure 3.2 Top and side views of the flat 2D hexagonal lattice. Gray balls are atoms with rods connecting nearest neighbors. The top view shows the sp^2 hybrids on each atom oriented to point to the nearest neighbors. The side view shows two p_z orbitals on neighboring atoms.

The overlapping of these in-plane hybrids along each interatomic axis leads to σ bonding and σ^* antibonding (i.e., symmetric and antisymmetric) combinations that are split by a large energy. If the bonding states are filled by one electron coming from each neighboring atom, then we have two-center bonding, that is, a textbook case of a pair of electrons shared by a pair of atoms. When intra-atomic overlap of hybrids is added, these bonding and antibonding states broaden into bands, but a gap still remains between these states. For the out-of-plane states, one has a single p_z orbital on each atom (see Fig. 3.2), which overlaps with those of its neighbors to form π and π^* bands. We will see that the sp^2+p_z basis will also help explain the electronic structure of other 2D boron sheets, but one will have to rotate the sp^2 basis to adapt it to the particular structure under consideration.

Theoretically, one could stabilize hexagonal boron by doping it with enough electrons to fill up all the bonding σ and π states to make it isoelectronic to graphene, but for a pure boron structure it is hard to imagine boron doping itself, so we need to consider other bonding geometries and lattices. Certainly, bulk phases of boron are not built from hexagons but instead from linked B_{12} icosahedra where triangular motifs abound and the B atoms have more than three neighbors (see Fig. 3.3).

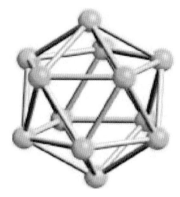

Figure 3.3   The B_{12} icosahedron underlying the stable bulk phases of B at ambient pressures. Various linking patterns of the icosahedra distinguish the phases.

Indeed, theoretical studies on pure boron molecules (clusters) showed that structures with B in triangular arrangements are quite stable and 2D B sheets based on a triangular lattice were proposed as stable structures [4–7, 21]. These triangular sheets are mechanically stable although their surface is not flat but prefers to buckle (i.e., some atoms move above the nominal sheet plane and some move below).

The electronic structure of triangular sheets is now discussed. Rather than showing band structures as a function of crystal momentum (as in Fig. 3.1) which is only of use for high-symmetry structures, it is more fruitful to coarse-grain the description and focus on the DOSs projected onto in-plane and out-of-plane states. The top panel of Fig. 3.4 displays the DOSs for the hexagonal structure that shows the separation of bonding and antibonding bands and that the Fermi energy cuts through the

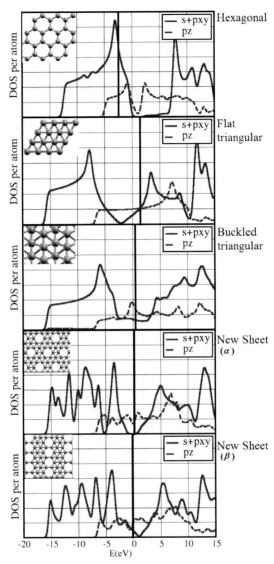

Figure 3.4   Densities of states (DOSs) for some 2D single-layered B sheets. Solid red curves are DOS-projected onto in-plane σ states (sum of s, p_x, p_y, and p_z projections) and blue dashed curves are projected onto out-of plane π states (p_z projection). The horizontal axis is energy in eV. The vertical thick black line marks the Fermi energy in each case. The insets in each case show top views of the atomic geometry. Reprinted with permission from Ref. [22].

bonding manifold. The second panel shows these DOSs for the flat triangular sheet (with an inset showing the geometry). The in-plane states no longer have a gap due to the larger number of neighbors and wider bands, but a zero in the in-plane DOS separates bonding σ and antibonding σ^* states; the out-of-plane states form a continuous band as expected for a simple structure with one p_z orbital per unit cell. The triangular lattice distorts in the out-of-plane direction to form a buckled structure shown one panel below: the changes in DOSs are small and also harder to interpret since we no longer have a reflection plane so projections onto in-plane and out-of-plane orbitals are not as informative.

At any rate, the triangular structure is mechanically stable and is thermodynamically more stable than the hexagonal lattice. As a prelude to the next section, we see that the Fermi level for the flat triangular sheet is not quite optimal either: it is somewhat too high and actually occupies some of the antibonding σ^* states. The consequences of this are discussed next, while the underlying reasons for the occurrence of buckling are dealt with further below.

3.3 Mixed-Phase Triangular-Hexagonal Boron Sheets

The situation regarding boron sheets changed significantly in 2007 and 2008 with three contemporaneous works. A DFT study of boron clusters identified stable boron fullerenes, the most notable being the B_{80} fullerene: a hollow spherical structure whose surface is composed from a mixture of pentagons and triangles. Its structure could be obtained from the carbon C_{60} fullerene by filling in the centers of the hexagons with B atoms to create triangular regions [13]. An examination of this curved surface and how it could be unfolded into a flat sheet revealed a previously unknown and highly stable 2D boron sheet structure, the "α sheet" described below [23]. A systematic examination focused on the extended sheet limit showed how an entire new class of single-layered boron sheets could be constructed that were more stable than the triangular sheet [22].

These sheets are composed of triangular and hexagonal regions mixed in various proportions, with the most stable structure being the α sheet. We will call this class of sheets mixed-phase triangular-hexagonal sheets as they smoothly interpolate between the triangular and hexagonal structures with a pronounced stability maximum.

These mixed-phase sheets can theoretically be constructed, as outlined in Fig. 3.5. One starts with a triangular sheet and chooses to remove a subset of atoms. A removal creates a hexagonal hole. The pattern of removals specifies a particular mixed-phase sheet. Clearly, the phase space of possible mixed-phase sheets is enormous. One parameter describing a removal pattern is the areal density of resulting hexagons or hexagon hole density η. This is defined as

$$\eta = \frac{\text{Number of resulting hexgonal holes}}{\text{Number of atoms in the original triangular sheet}}$$

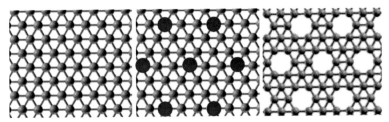

Figure 3.5 Schematic construction of mixed-phase B sheets. Left: Start with a triangular sheet; this is a top view with gray balls being B atoms and nearest neighbors connected by rods. Center: Select some subset of atoms (red circles) and remove those atoms. Right: The resulting structure is a mixed-phase triangular-hexagonal sheet, in this particular case the α sheet. The specific pattern of atomic removals specifies the resulting mixed-phase sheet.

Thus $\eta = 0$ specifies the pure triangular sheet, while the hexagonal sheet is a sheet with $\eta = 1/3$. Of course, for a given η, the pattern of removals may be important and this makes for the huge phase space. Luckily, first-principles calculations show that the most thermodynamically stable sheets have the hexagons spaced apart and as evenly as possible and that the opposite extreme of aligning the hexagonal holes in lines generally produces less stable sheets overall [22, 24]. Therefore, the primary focus of the subfield has been on B sheets with evenly spaced hexagons.

For the reasons discussed immediately below, while the even hexagon distribution may be the ground state, actual experimental realizations will most likely present much more complex structures

Figure 3.6 displays the binding energy per B atom for single-layered 2D B sheets versus η. The binding energy is defined as the energy of an isolated spin-polarized B atom minus the energy per atom of a given sheet; more positive binding energies mean more stable structures. We see a maximum in binding at $\eta = 1/9$, and the most optimal structure corresponds to the α sheet (Figs. 3.5 and 3.7). The binding energies are plotted separately for buckled structures and structures that are forced to be flat, and there is an interesting asymmetry in that only sheets with $\eta < 1/9$ buckle to lower their energy. The buckling is of most importance for small η sheets (i.e., large triangle density) and becomes much less significant as the stability maximum at $\eta = 1/9$ is approached.

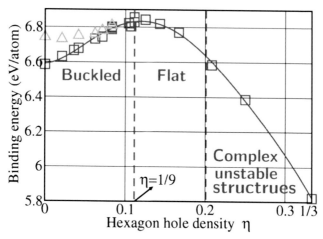

Figure 3.6   DFT binding energy of single-layered 2D boron sheets versus η with evenly spaced hexagonal holes. Squares are for structures optimized under the constraint of being flat with no out-of-plane buckling. Triangles are for structures where buckling is permitted. Reprinted with permission from Ref. [24].

A number of sheets structures are illustrated in Fig. 3.7. The somewhat arbitrary naming convention uses a letter followed by the η value of the sheet; T is reserved for the $\eta = 0$ triangular sheet, A is reserved for the optimal $\eta = 1/9$ α sheet, and letters are

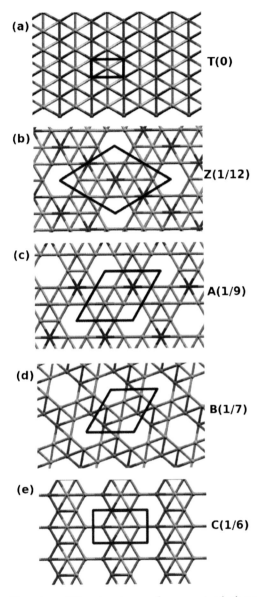

Figure 3.7 Top view of the structures of some mixed-phase sheets. Red lines show the primitive cells. The buckling patterns are indicated by coloring: out-of-plane coordinates increase as color changes from green to gray to purple (e.g., green means negative z coordinates, gray means $z \approx 0$, and blue means $z > 0$). Reprinted with permission from Ref. [24].

monotonically "increased" (with period wrapping through Z) with increased η. The figure shows the buckling pattern of each sheet when it happens spontaneously to lower the energy, as well as the buckling pattern that one can force on an otherwise flat sheet by imposing uniform in-plan compressive strain.

Returning to the binding energy plot in Fig. 3.5, the simplest thermodynamic interpretation of the plot is that the most stable sheets are flat and have the α sheet structure. However, some consideration of the plot leads one to conclude that this is quite unlikely in practice because the maximum is rather flat versus η. Namely, any actual experimental growth of these structures at finite temperature will invariably result in complex and disordered structures with η in the neighborhood of 1/9. This is because the energy close to the maximum depends very weakly on η but the configurational entropy gained by disordering the hexagonal holes, both in terms of position and in terms of density η, is probably unavoidable. Any growth of pure B sheets will have to be kinetically limited because 2D sheets of B are not the ground-state structure of boron (the ground-state structures are the 3D phases built from linked icosahedra). The kinetic limitations necessary to grow the structure will then most likely lead to significant entropy in the realized sheet. A more careful theoretical examination of disordered sheets and their resulting electronic structure and transport properties is an open question in the field as is the most likely structures realized in an actual experiment. These considerations also carry over to the structure of pure B nanotubes that are theoretically constructed from wrapping extended 2D sheets.

3.4   Stability of Mixed-Phase Boron Sheets

In this section we describe, from two separate viewpoints, the stability of the mixed-phase sheets and the dependence of the binding energy on the hexagon hole density η. The two viewpoints are closely related but focus on different ways of analyzing the electronic structure of the sheets.

A first viewpoint is based directly on the electronic band structure of the sheets and their fillings. The starting point is that a stable structure will fill the low-energy bonding states while leaving empty the high-energy antibonding states. Furthermore, in-plane and out-of-plane states are weighed differently: as exemplified by Fig. 3.2, the in-plane sp^2 hybrids are quite directional and their large overlaps lead to strong splittings of bonding and antibonding states, whereas p_z orbitals have much weaker overlaps. Therefore, to the lowest order, the most stabilizing approach is to focus on filling all σ bands, while leaving empty all σ^* bands, while filling in low-energy π bands is secondary. The next step is to try to understand the DOS of the triangular sheet in the same way we analyzed that of the hexagonal sheet. In the triangular structure, each B atom has six nearest neighbors but only three valence electrons, so a two-center bonding scheme where pairs of atoms share electron pairs is difficult to justify. The answer instead is to shift the center of the bonding away from the interatomic axis and view the triangular lattice as a set of triangles where the electrons are shared about the center of the triangle with the three atoms at the corners. This is denoted as three-center bonding in the chemical literature [25]. As a confirmation, first-principles calculations had qualitatively noted the concentration of electrons in the triangular centers [26].

We now describe, in detail, how the three-center bonding is achieved and the structure of the bands for the triangular lattice. We rotate the sp^2 hybrids on each atom to point to the center of the triangles, as shown in Fig. 3.8. Within each equilateral triangle the hybrids overlap, and the D_3 symmetry ensures enforces a low-energy bonding state (b) and a doubly degenerate antibonding pair (a*). These then broaden into bands due to intertriangle hopping elements. Looking at the DOS in Fig. 3.4, we do indeed see that the in-plane DOS has a zero, which is the separation between the b-derived and a*-derived states. The out-of-plane p_z derived π states also form a continuous band. One would like to populate the b-derived band with two electrons, leave the a*-derived bands empty, and dump the remaining electron into the π band. However, there is no guarantee that the energies of the atomic states for boron will make this happen: after all, from elementary tight-binding considerations, the center of the π band is the atomic eigenvalue of the p_z state, while the center of the b-derived band is related to the energy of the sp^2 hybrids which depends on the s-p splitting; there is no reason for the *s*-*p* splitting to have the desired value. And indeed, as Fig. 3.4 shows, the Fermi level is slightly off from the ideal position and fills some of the antibonding a*-derived bands.

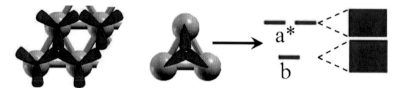

Figure 3.8   Schematic showing the orientation of sp^2 hybrids and three-center bonding in flat triangular boron sheets. Left: The sp^2 hybrids are chosen to point to the center of the triangles. Center and right: Within each triangle, the overlap of the hybrids gives rise to a low-energy bonding state b and doubly degenerate antibonding states a*. These states then broaden into bands due to intertriangle hopping matrix elements. Reprinted with permission from Ref. [22].

From an electron-doping perspective, the hexagonal lattice is an electron acceptor, since taking electrons into its unoccupied low-energy in-plane states stabilizes it. Conversely, the triangular sheet is an electron donor since it prefers to lose the high-energy electrons in its antibonding states. In this way, one can rationalize the stability maximum versus η in Fig. 3.6. By mixing hexagons and triangles, one is in some sense self-doping the system toward optimal stability. And as shown in Fig. 3.4, the most optimal α sheet has that triangle-hexagon mixture, which puts the Fermi level at the zero of the in-plane DOS. Sheets that are nearby perturbations of the α sheet (e.g., the β sheet in Fig. 3.4) have the Fermi level almost but not quite at the optimum position. In this way, the variable stability of the mixed-phase 2D sheets is understood as an interplay between two-center hexagonal and three-center triangular regions that dope and electronically compensate each other. The happiest marriage of these two phases happens for the α sheet. What is quite unusual in this representation is that we have a homopolar material that is self-doping with no obvious point defects in sight.

The second viewpoint on the stability of the mixed-phase sheets has us trying to simplify the description of the electronic structure away from the band structure and DOS and to try to create a very simple electron-counting scheme. This is in fact possible and one ends up with some unusual and interesting observations. The starting point is to try to formalize the above ideas of in-plane and out-of-plane states and count the number of such states starting from low energy and up the "separation energy," defined as the zero in the in-plane DOS. That the in-plane DOS should have a zero separating bonding from antibonding states for a large variety of 2D boron sheets is not obvious a priori, but it is verified by examination of the electronic structure of a number of sheets [27] as well as by the more intuitive observation that both the hexagonal and triangular sheets have this property and the mixed-phase sheets are an interpolation between the two.

One can use DFT DOS calculations to extract the number of in-plane N_σ and out-of-plane states N_π (per spin) for boron sheets up to the separation energy for a wide range of η. Being extensive quantities, one normalizes them by M, the number of atoms in the original triangular lattice (which generated the sheet in question by removal of atoms to generate hexagon whole density η). As Fig. 3.9 shows, an unexpected result is found: N_σ/M is completely flat and equal to unity as a function of η, while N_π/M hovers very close to 1/3. Why this is the case is instructive and is explained below, but for now we take these results for granted and make use of them to understand the stability of the α sheet. Namely, as argued above, the most stable structure has the Fermi level right at the separation point. Thus with two spin-paired electrons filling each state, we must have that $2(N_\sigma + N_\pi)$ equals the number of electrons. The number of electrons is $3M(1 - \eta)$ because each boron atom has three valence electrons, and we have removed a fraction η of the atoms by drilling hexagonal holes. So $2(N_\sigma + N_\pi)$ = $3M(1 - \eta)$ and substituting $N_\sigma = M$ and $N_\pi = M/3$ immediately yields $\eta = 1/9$, which is the α sheet.

But why do N_σ and N_π behave in this unusual manner? After all, changing η changes the areal density of B atoms as well as their connectivity (six-neighbored triangular regions to three-neighbored hexagonal regions): how can the number of in-plane bonding states remain fixed? To answer this question, we will need an ab initio localized approach for describing the electronic structure, namely maximally localized Wannier functions (MLWFs) [28–30]. In brief, the MLWF prescription delivers localized and orthonormal functions that span exactly the same Hilbert space as the eigenstates in some desired energy range. Thus they form a complete and orthonormal ab initio localized basis. In cases where one already has information about the electronic structure and is instead focused on understanding the chemical bonding, for example, in our case we know the energy range for the bonding states due to the zero in the in-plane DOS, one can use them to confidently describe the electronic structure with no approximations. Here, we focus on MLWFs generated to describe the bonding manifold of the boron sheets (i.e., the occupied valence states up to the separation energy).

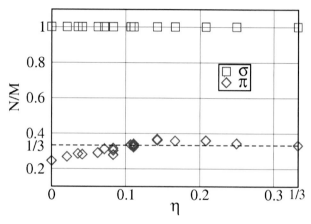

Figure 3.9   Number of in-plane states (σ, red squares) and out-of-plane states (π, blue diamonds) per spin up to the separation energy as a function of η. Reprinted with permission from Ref. [27].

We begin with the triangular sheet. Figure 3.10 shows that for the in-plane bonding states, two nonequivalent sets of MLWFs can be found depending on how the localization procedures is seeded. The first set shown in Fig. 3.10a emphasizes the triangular nature of the in-plane network as described above. These bonding MLWFs are centered in the middle of the triangles as expected. Filling the bonding bands up to the separation energy is equivalent to putting two spin-paired electrons into each such MLWFs. However, a second set of MLWFs can be generated that are centered on the midpoint between the boron atoms, as shown in Fig. 3.10b. The layout of the MLWF centers is that of a hexagonal lattice, which is a critical observation for what follows below. Despite the obvious difference between the two sets of MLWFs in Figs. 3.10a and 3.10b, we emphasize that both orthonormal sets span the Hilbert space of bonding electronic states of these boron sheets exactly and equally well so that they are equivalent descriptions of the same electronic structure (i.e., unitary transforms of each other). This simply emphasizes the point that there is no unique way to partition delocalized electrons in an extended structure into localized states. Different decompositions emphasize different aspects of the electronic structure, and some are more effective than others at explaining a particular phenomenon.

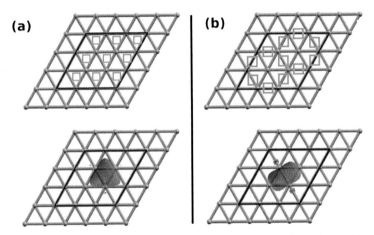

Figure 3.10 In-plane σ MLWFs for the flat triangular sheet. Two nonequivalent sets of MLWFs can be generated by the MLWF localization procedure. (a) MLWF centered on the triangles. Green squares in the top figure locate the centers of the MLWF, while the bottom figure shows an isosurface of the MLWF (red are positive and blue are negative values). (b) MLWF centered between pairs of atoms. Reprinted with permission from Ref. [27].

For what follows, the set of MLWFs of the type in Fig. 3.10b are most fruitful in providing a simple picture. One can generate these types of MLWFs by appropriate seeding of the localization procedure for a variety of 2D boron sheets. Figure 3.11 shows the resulting MLWFs for a range of η values. The most striking feature of these figures is that the number and centers of the MWLFs do not budge from their hexagonal H(1/3) configuration. Of course, the actual microscopic shape of the MLWFs do change with η: the

MLWFs adjacent to triangular regions tend to spread out slightly into the triangular regions; and of course, both the on-site and hopping elements between these MLWFs depend strongly on η since the potential is changing dramatically (i.e., the band widths and average band energies evolve with η). However, the key point here is that the number and layout is invariant. Thus, if we view these boron sheets as starting with hexagonal holes that are then filled with atoms to form triangular regions, the added atoms don't actually change the basic aspects of the rigid in-plane bonding network. A similar analysis for the out-of-plane π states shows that their number and centers are also invariant [27]. Thus a fixed number of localized orbitals with fixed central locations describe the band structure for variable η.

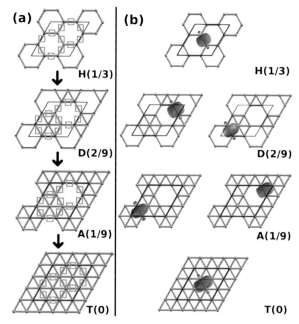

Figure 3.11 In-plane MLWFs for flat boron sheets starting with the hexagonal H(1/3) and ending with the triangular T(0) sheets. Green squares in (a) mark the center of the MLWFs for each sheet, while isosurfaces in (b) show the nonequivalent MLWFs for that sheet. Other MLWFs are generated by translations or symmetry operations. Red solid lines show unit cells used to generate the MLWFs. Reprinted with permission from Ref. [27].

The implication is that when we add a boron atom to a hexagonal hole to increase η, we can say that the added boron doesn't actually contribute new localized states to the bonding manifold but instead simply releases its three valence electrons into the available bands. This unusual behavior where the added boron atoms perfectly dope the system is called self-doping [27]. Compared to the two-center/three-center description above, this is an alternate view of bonding in boron sheets that focuses primarily on electron count and an underlying rigid set of localized orbitals that give rise to the bonding manifold. In practice, the self-doing picture is of enormous help in trying to get a simple zeroth-order view of more complex boron nanostructures involving metal atoms where charge transfer from metal to boron occurs: for example, the self-doping approach allows one to quickly zoom in on the η values of the most stable boron sublattice as the fraction of metal atoms and the electron transfer to the boron is varied [27]. This greatly simplifies the search for the ground-state structure of these more complex nanomaterials.

3.5 Buckling of Boron Sheets

Section 3.2 explained that some 2D boron sheets have a buckled or "puckered" ground state where some B atoms move above, and some below, the nominal plane of the sheet; namely, the flat structure for these sheets is either mechanically unstable or a higher-energy metastable state [2–5, 8–12]. However, only sheets with $\eta < 1/9$ prefer to buckle, while those with $\eta > 1/9$ are flat in their ground state. Why do some sheets prefer to buckle but not others? As the next section on double-layered B sheets shows, the answer to this question has ramifications beyond simply understanding the properties of single-layered B sheets.

Since all known single-layered boron sheets are metallic and most, excluding those very near the α sheet, have both in-plane σ and out-of-plane π states at the Fermi level, one might loosely rationalize this behavior as a type of Peierls distortion. The σ and π states, which are forbidden to interact for flat sheets due to parity since the sheet plane is a mirror plane, can mix or hybridize due to the reduced symmetry, and this mixing lowers the energies of the occupied states especially around the Fermi level. For boron nanotubes built from rolling up the α sheet, the surfaces of the nanotubes are found to be buckled and the electronic density shows clear signs of σ-π hybridization [23, 31]. However, is the σ-π hybrization the driving cause for buckling or a by-product? Once the sheet symmetry is reduced by the buckling, σ-π mixing is symmetry-allowed and will appear generically.

Our present understanding is that an explanation based σ-π mixing as the causal driving force for buckling is untenable. A priori, such an explanation is difficult to reconcile with the asymmetry of buckling with η about 1/9: all sheets with η away from 1/9 have both σ and π states available at the Fermi energy for potential mixing but only those with $\eta < 1/9$ actually buckle. A posteriori, one can analyze the first principles results by explicit separation of the band energy E_{band} out of the total ground-state energy E_{tot}:

$$E_{\text{tot}} = E_{\text{band}} + E_{\text{rep}}$$

Here, E_{band} is the sum of all the occupied Kohn–Sham eigenvalues, while E_{rep}, called the repulsive energy, is whatever must be added to the band energy to recover the total energy. The choice of the word "repulsive" is based on the fact that in almost all tight-binding approaches to electronic structure, the total energy is decomposed in this very manner, with E_{rep} approximated by a sum of interatomic repulsive terms, but we emphasize that in a first-principles approach there are no such approximations. An explanation that posits σ-π hybridization as the cause for the lowering of the total energy requires that it E_{band} is reduced by buckling. Unfortunately, the actual ab initio energies do not support this hypothesis [24]: the band energy generally increases due to buckling, and no systematic trends in the changes of band energy or repulsive energy due to buckling are found versus η.

Clearly, we need an alternative physical picture and associated decomposition of the total energy. Another explanation for buckling stems from the intuitive idea that buckling might increase the surface area of a sheet (i.e., same number of atoms but larger interatomic distances). With a fixed number of atoms and associated electrons, a larger surface area implies a lower areal density of electrons, and one can speculate that it is the lowering of electron density that drives the buckling. While intuitively plausible, attempting to define the surface area of a structure at the atomic scale is difficult and ambiguous. However, this picture argues that one should view the electrons in the boron sheet as forming a type of 2D electron gas whose energetics is largely dominated by their average electron density. One would expect that lowering the average electron density would lower the electronic kinetic energy. Thus, one is led to decompose the total energy in a way appropriate for a quantum electron gas:

$$E_{tot} = E_{kin} + E_{es} + E_{xc}$$

Here E_{kin} is the total kinetic energy of the electrons, E_{es} is the total classical electrostatic interaction of all charge densities (electron–electron, electron–ion, and ion–ion), and E_{xc} is the total exchange and correlation energy of the electrons. In Fig. 3.12, we show the dependence of each of these energies for flat boron sheets as a function of η. We see a monotonic dependence of each energy term versus η. The trends are indeed as expected: increasing η decreases the areal density of boron atoms and thus the areal electron density; the kinetic energy of an electron gas decreases with decreased density due to the lower Fermi momentum or Fermi energy; and the total electrostatic interactions of all charges must rise with decreasing density since net neutral atoms are having a larger average separation and interelectron exchange and correlation effects stabilize high electron densities. With this decomposition, one can examine how each energy term changes when the boron sheet buckles. One finds that for any boron sheet, and not only those with $\eta < 1/9$, E_{kin} decreases with buckling while E_{es} and E_{xc} both increase [24]. This confirms that it is the reduction of kinetic energy accompanying buckling that drives the lowering of total energy; however only for high areal electron densities (i.e., $\eta < 1/9$) is the lowering of E_{kin} sufficient to overcome the raising of $E_{es} + E_{xc}$.

Trying to establish a direct relation between buckling and reduction of average electron density is more problematic as the latter is very hard to define. One possible definition is to compute an electron density $\bar{n}$ (using the electron density itself as a weighing function). Using this definition of $\bar{n}$, one can show $\bar{n}$ is monotonically related to η and that buckling does indeed lower $\bar{n}$ [24], which closes the loop in the argument. However, the lack of a good definition of average electron density is unsatisfactory and an open problem.

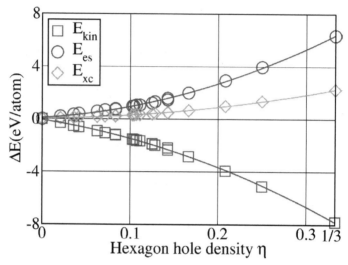

Figure 3.12 Kinetic, electrostatic, and exchange-correlation energies versus η for flat 2D boron sheets. All three energies are plotted with respect to their respective values at $\eta = 0$ (zero of energies). The symbols are calculated DFT results. Solid curves are guides for the eye. Reprinted with permission from Ref. [24].

We now summarize our present understanding of the reasons underlying the buckling of boron sheets. The driving force for buckling is the electronic kinetic energy, and this kinetic energy is always lowered by buckling. The relation of electron density to buckling is more complex: Buckling reduces the average electron density in the boron sheet, and the average electron density is related one-to-one to η so that a decreased average electron density corresponds to increased η. Therefore, since the binding energy curve for flat boron sheets versus η (Fig. 3.6) has a maximum at $\eta = 1/9$, sheets with η below 1/9 can lower their total energy by buckling since that effectively increases their η value, while sheets starting with η above 1/9 cannot lower their total energy by buckling (and can only raise it). We will this knowledge immediately below when considering multilayered boron sheets.

3.6 Double-Layered Boron Sheets

The aim of this section is to describe what stable boron sheets may look like when they are allowed to be more than a single atom thick. We will be restricting the discussion to the simplest case of a bilayer of boron (i.e., a boron sheets two atoms tall).

Before beginning the discussion, we make some comments on how such theoretical results are best understood in relation to experiment. Carbon in the form of graphene provides a good example. Generally, a single-layered system will not be as stable as a multilayer due to attractions between the layers. In the case of graphene, we know that graphite, composed of stacked sheets of graphene held together by van der Waals interactions, is more stable than graphene. In fact, for carbon one is lucky that graphite is the ground state when compared to the three-dimensionally bonded diamond structure: the material prefers to be in a sheet-like form, which explains the great success in growing atomically thin carbon structures (fullerenes, nanotubes, and graphene). Theoretically, if one blindly searches for the most thermodynamically stable structure for carbon, one ends up with graphite; however, control over the growth and kinetics of carbon is able to deliver with high specificity either clusters (fullerenes) or single-walled or multiwalled nanotubes. In the same way, double-layered boron sheets will be more stable than single-layered sheets, and thick bulk-like structures containing B_{12} icosahedra are even more stable than both. One envisions that the growth conditions for boron may one day be controlled in such a way as to deliver metastable structures such as single or few-layered boron sheets.

3.7 Interlayer Bonding in Double-Layered Boron Sheets

Pioneering work on boron sheets noted the presence of strong interlayer boron–boron bonds between stacked sheet-like structures and the stabilization of double-layered structures [6, 7, 32]. To study this type of sysem systematically from a theoretical point of view, one can consider taking two identical single-layered boron sheets, aligning them to be parallel, and bringing them together to see what happens.

The result is dependent on η [24], as illustrated by Fig. 3.13. For a pair of $\eta < 1/9$ sheets, which separately prefer to buckle, interlayer bonds form between buckled atoms that have moved into the space between the two sheets. The resulting bond lengths are about 1.7 Å, a value quite representative of B–B bonds in single-layered sheets. Hence, one can classify the bonds as being

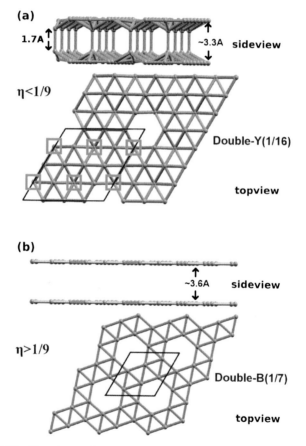

Figure 3.13 Structures of two representative double-layered boron sheets viewed from the top and side. (a) A pair of $\eta < 1/9$ form interlayer bonds. The top view shows the unit cell (red lines) and the atoms that form interlayer bonds (green squares). (b) A pair of $\eta > 1/9$ sheets remains flat and relatively far apart. Reprinted with permission from Ref. [24].

genuine chemical bonds. However, for a pair of $\eta > 1/9$ sheets, both sheets stay essentially flat and are about 3.5 to 3.7 Å apart, which classifies their interaction as being weak and of van der Waals type. The binding energy is about an order of magnitude smaller than a chemically bound pair of $\eta < 1/9$ sheets.

This behavior is completely unsurprising given the tendencies of the individual sheets toward buckling. Sheets that prefer to buckle in isolation use the opportunity to buckle and make new bonds with each other. This type of study gives useful guidelines for stabilizing boron bilayers or multilayers but is not conclusive or exhaustive. Two pairs of identical sheets make the theoretical analysis straightforward and, in addition, are lattice-matched so that the corresponding buckling atoms from both sheets are in perfect registry to form interlayer bonds. But, in principle, two different sheets with $\eta < 1/9$ can form a number of interlayer bonds (despite the mismatched registry). Whether pairs of mismatched but identical bonded sheets can be more stable than the pair of identical sheets is an open question.

The most stable double layer constructed using the above procedure is made of a pair of $\eta = 1/12$ sheets, and its structure is shown in Fig. 3.14. The energetic cost of moving off $\eta = 1/9$ has been balanced by the energy reduction from buckling and forming interlayer bonds. Interestingly, this double-layered sheet is predicted to be insulating with a band gap of 0.8 eV in DFT [24]. Speaking figuratively, it would appear that the formation of the interlayer bonds has devoured the electrons about the Fermi level in each sheet. Therefore, while single-layered boron sheets are always found to be metallic, stable double-layered (and possibly multilayered) boron structures may in fact be insulating—which is also true of bulk boron.

3.8 Outlook

This chapter has summarized current understanding on the structure, energetics, stability, and electronic structure of 2D boron sheets both atomically thin and two atoms thick. As is obvious to the reader, the discussion has been purely theoretical because to date no experiments have realized extended 2D boron sheets much

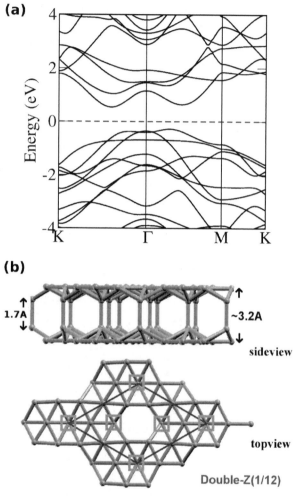

(a)

(b)

sideview

topview

Double-Z(1/12)

Figure 3.14 The most stable double-layered boron sheet constructed from two identical single-layered η = 1/12 sheets. Side and top views are shown. This sheet is an insulator. Reprinted with permission from Ref. [24].

less characterize them. In fact, in this author's opinion, the major challenge facing the entire field of study of boron nanostructures is the heavy domination by first-principles theory. Clearly, achieving the growth of boron sheets or nanotubes and fullerenes with atomically thin walls is much more challenging than their carbon counterparts. This is at least in part due to the fact that, unlike carbon, boron in its bulk ground state prefers a 3D crystalline structure composed of linked icosahedra rather than a structure composed of layered 2D motifs.

Focusing on 2D boron sheets, an outstanding question is to understand on what substrates the growth of boron in sheet form is kinetically viable. This will require both experimental and theoretical studies on the diffusion barriers for B on the substrate, a good lattice match of the sheets to the substrate, the conditions favoring formation of 2D islands versus 3D clusters, the temperature ranges for which the boron atoms stay on the substrate rather than either diffusing into the substrate or chemically reacting with the surface to form a disordered alloy, etc. One possible class of substrates to consider is transition metals, which can present ordered and relatively flat surfaces. These substrates may bind the boron atoms strongly enough to force the boron to adopt a 2D structure at low coverage. In addition, being metallic, one can easily image the resulting surface structures with methods such as scanning tunneling microscopy to reveal the atomic-scale structure. Of course, electron transfer from the metal to the boron sheet will change its most stable η value in a predictable manner [27], depending on the substrate chosen. This might actually enlarge the types of boron sheets attainable via this proposed growth method.

References

1. Novoselov K. S., Geim A. K., Morozov S. V., Jiang D., Zhang Y., Dubonos S. V., Grigorieva I. V., and Firsov A. A., *Science*, **306**, 666 (2004).

2. Boustani I., *Chemical Physics Letters*, **233**, 273 (1995a).

3. Boustani I., *Chemical Physics Letters*, **240**, 135 (1995b).

4. Boustani I., *Physical Review B*, **55**, 16426 (1997a).

5. Boustani I., *Surface Science*, **370**, 355 (1997b).

6. Boustani I., Quandt A., Hernandez E., and Rubio A., *Journal of Chemical Physics,* **110**, 3176 (1999a).

7. Boustani I., Rubio A., and Alonso J. A., *Chemical Physics Letters* **311**, 21 (1999b).

8. Zhai H. J., Wang L. S., Alexandrova A. N., and Boldyrev A. I., *Journal of Chemical Physics,* **117**, 7917 (2002).

9. Zhai H. J., Kiran B., Li J. L., and Wang L. S., *Nature Materials*, **2**, 827 (2003a).

10. Zhai H. J., Alexandrova A. N., Birch K. A., Boldyrev A. I., and Wang L.-S., *Angewandte Chemie International Edition*, **42**, 6004 (2003b).

11. Alexandrova N., Boldyrev A. I., Zhai H. J., and Wang L. S., *Journal of Physical Chemistry A*, **108**, 3509 (2004).

12. Kiran, Bulusu S., Zhai H. J., Yoo S., Zeng X. C., and Wang L. S., *Proceedings of the National Academy of Sciences of the U.S.A.* **102**, 961 (2005).

13. Gonzalez Szwacki N., Sadrzadeh A., and Yakobson B. I., *Physical Review Letters*, **98**, 166804 (2007).

14. Gonzalez Szwacki N., *Nanoscale Research Letters*, **3**, 49 (2007).

15. Zope R., *Europhysics Letters*, **85**, 68005 (2009).

16. Özdoğan, Mukhopadhyay S., Hayami W., Güvenç Z. B., Pandey R., and Boustani I., *Journal of Physical Chemistry C*, **114**, 4362 (2010).

17. Prasad L. V. K., and Jemmis E. D., *Physical Review Letters*, **100**, 165504 (2008).

18. Zhao J., Wang L., Li F., and Chen Z., *Journal of Physical Chemistry A*, **114**, 9969 (2010).

19. De S., Willand A., Amsler M., Pochet P., Genovese L., and Goedecker S., *Physical Review Letters*, **106**, 225502 (2011).

20. Evans M. H., Joannopoulos J. D., and Pantelides S. T., *Physical Review B*, **72**, 045434 (2005).

21. Chacko S., Kanhere D. G., and Boustani I., *Physical Review B*, **68**, 035414 (2003).

22. Tang H., and Ismail-Beigi S., *Physical Review Letters*, **99**, 115501 (2007).

23. Yang X., Ding Y., and Ni J., *Physical Review B*, **77**, 041402(R) (2008).

24. Tang H., and Ismail-Beigi S., *Physical Review B*, **82**, 115412 (2010).

25. Durrant P. J., and Durrant B., *Introduction to Advanced Inorganic Chemistry* (Wiley and Sons, New York, 1962).

26. Kunstmann J., and Quandt A., *Physical Review B*, **74**, 035413 (2006).

27. Tang H., and Ismail-Beigi S., *Physical Review B*, **80**, 134113 (2009).

28. Marzari N., and Vanderbilt D., *Physical Review B*, **56**, 12847 (1997).

29. Souza I., Marzari N., and Vanderbilt D., *Physical Review B*, **65**, 035109 (2001).

30. Mostofi, Yates J. R., Lee Y.-S., Souza I., Vanderbilt D., and Marzari N., *Computer Physics Communications*, **178**, 685 (2008).

31. Singh K., Sadrzadeh A., and Yakobson B. I., *Nano Letters*, **8**, 1314 (2008).

32. Sebetci, Mete E., and Boustani I., *Journal of Physics and Chemistry of Solids*, **69**, 2004 (2008).

Chapter 4

Ab initio Study of Single-Walled Boron Nanotubes

Kah Chun Lau and Ravindra Pandey

Physics Department, Michigan Technological University, Houghton MI 49931, USA

kclau@mtu.edu, pandey@mtu.edu

4.1 Introduction

For future nanotechnology and molecular electronic devices, the reduced dimension of carbon nanostructures (e.g., carbon nanotubes [CNTs] [1] and graphene [2]) is one of the great interesting candidates to the scientific community since the discovery of carbon fullerenes [3]. Despite having similar local atomic configurations, fullerenes, graphenes, and nanotubes can exhibit different electronic and mechanical properties based on their distinct physical dimensionality and unique periodic boundary condition. Despite their promising potentials in device application, the future developments of nanotechnology devices and architectures remain uncertain and cannot be realized with carbon allotropes only. Therefore the quest for the novel materials has never stopped. Stimulated by CNTs, there are also other types

Handbook of Boron Nanostructures
Edited by Sumit Saxena
Copyright © 2016 Pan Stanford Publishing Pte. Ltd.
ISBN 978-981-4613-94-1 (Hardcover), 978-981-4613-95-8 (eBook)
www.panstanford.com

of elemental or compound nanotubes been synthesized since then. The reported elemental nanotubes can be of a pure element, Au [4, 5], Bi [6], Si [7, 8], etc., whereas for compound nanotubes, BC_3 [9, 10], BN [11, 12], MoS_2 [13, 14], etc., have been reported. Among the recently reported elemental nanotubes, the nanotubes that consist of the boron atom (i.e., boron nanotubes, BNTs) are relatively less conventional and exotic and remain elusive relative to CNTs in the scientific community because they are technically more challenging to synthesize.

Compared with the other quasi 1D boron nanostructures, for example, nanowires, nanoribbons, and nanowhiskers [15–17], BNTs can be categorized as a new class of topological structure in boron. While the boron nanowires, nanoribbons, and nanobelts are all found to be bulk-like (i.e., either in crystalline or in amorphous phase) [15–17], details of the structural morphology of BNTs remain to be verified and characterized experimentally. For BNTs, it has been proposed and theoretically studied extensively [18–33] since the last decade; however, very limited experimental reports on BNTs synthesis can be found in literature [34, 35]. To date, the growth of single-walled crystalline BNTs in the laboratory remains a challenge in this field. According to the first reported crystalline BNTs synthesized on the basis of the thermal evaporation method [35], the actual atomic configuration and chirality of the individual BNT is unclear and remains to be determined experimentally. As reported [34–35], all BNTs have diameters in a range from 10 to 40 nm and are multiwalled. The purity of the synthesized BNTs remains to be improved, and the percentage yield of boron nanowires (BNWs) that are mixed with the BNTs in the sample has to be reduced. Therefore to explore the basic properties of BNTs, the ab initio (or first-principles) simulation of BNTs can be extremely useful. To further understand the basic features of this boron nanostructure, a fundamental understanding of the elemental boron bulk crystalline phase, which is thermodynamically more favorable in nature, is needed in order to reveal the basic chemical bonding of boron atoms. Assuming the understanding in bulk crystalline phases of elemental boron is established, ideally their reduced dimensionality features at the nanostructures (e.g., BNTs) can be deduced.

As a neighbor of carbon, which is the fifth element in the periodic table, boron occupies an interesting transitional position between nonmetallic and metal elements. The interest in boron has always been high due to its "electron deficient" (i.e., the number of available valence electrons, $2s^2 2p^1$, is less than the available orbitals in the electronic configuration of atomic valence shell) nature [36, 37]. The consequences for the nature of the chemical bonds of electron-deficient materials are well-known, and they may be summarized as follows. First, the ligancy of electron-deficient materials is higher than the number of atoms and even higher than the number of stable valence orbitals. Second, electron-deficient materials cause adjacent atoms to increase their ligancy to the values larger than the orbital numbers [38]. In general, the electron-deficient nature does not suggest that it is inferior in bonding but simply that novel complex structures based on elemental boron are expected to be adopted in nature. Therefore for nanoscopic-size boron clusters, their potential energy surfaces or energy landscapes are typically found to be flat and "glasslike" with enormous density of similar local minima, in contrast to carbon and boron nitride systems of similar sizes [39]. Thus, the complex polymorphic feature is therefore expected to be dominant in boron nanostructures [39–42], analogous to elemental boron solids.

With insufficient electrons to support a structure, this unique electron-deficient electronic bonding character generally led to the peculiar hybridization of s- and p-orbital in bonding character, that is, the coexistence of localized two electron two-center (2-c) and delocalized two-electron three-center (3-c) bonding to resolve its electron deficiency (Fig. 4.1). Thus boron compounds and elemental boron solids exist in a variety of intriguing and peculiar geometric configurations in topologically connected complex atomic networks that render polymorphism in bulk crystalline phases [43–45], in contrast to the well-defined carbon system. There are reports of numerous allotropes of boron; however, only few of them are comparatively well characterized [43–45]. Under ambient pressure, the morphology of elemental boron solid is generally stabilized by the 2-c backbone covalent bonding in hexagonal lattice and the 3-c delocalized bonding that spreads in a triangular lattice, which utilized the B_{12} icosahedrons as basic units. For α-B_{12} rhombohedra solids, the 2-c and 3-c bonds can be typically found in intra- and intericosahedral chemical bonds within the solid, as shown in Fig. 4.1. Depending on a broad range of possible local arrangements of boron atoms, the interplay of 2-c and 3-c bonds often leads to the emergence of polymorphs of elemental boron structures both in crystalline bulks and in nanostructures [20, 28, 44] (Fig. 4.1). With this pronounced polymorphism in nature, it is reasonable to assume that not only one but multiple BNT structural motifs can be possibly realized in nature.

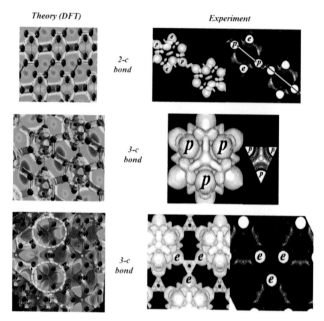

Figure 4.1   The localized two-electron two-center (2-c) (i.e., interico-sahedral bond) and the delocalized two-electron three-center (3-c) bond (i.e., intraicosahedral and intericosahedral bonds) that typically are found in elemental boron bulk crystalline phases, for example, α-B_{12} icosahedral bulk solid found in density functional theory (DFT) [20], and experimental electron density distribution obtained by using the maximum entropy method with synchrotron radiation powder data [44]. The "p" and "e" are referring to "polar" and "equatorial" atoms in the B_{12} icosahedron, as defined in Ref. [44].

4.2   The Basic Construction Model of Single-Walled Boron Nanotubes

For boron atoms in a nanotube configuration, its simplest form is the single-walled boron nanotube (SWBNT). All the boron atoms that are constrained in 1D tubular geometry generally experience positive strain energy along the circumference, depending on the curvature (or a diameter) for an SWBNT according to elasticity theory. Thus following this theory, a 1D SWBNT with a smaller diameter (i.e., bigger curvature) generally should be energetically less favorable or thermodynamic less stable than a bigger-diameter (i.e., smaller-curvature) SWBNT. Following this simple argument, so that for large-diameter nanotubes with vanishing curvature, we are assured that surface structure of the most stable single-walled nanotube will be asymptotic equivalent to a single-layer 2D boron sheet. Thus within the simplest approximation, an SWBNT can be considered as a rolled boron sheet, depending on its diameters and chiral vectors. However, the atomic structure of SWBNT is still unclear because the single-layer quasi-2D boron sheet that constitutes the walls of BNTs has not yet been experimentally discovered or characterized. So to understand the basic properties of an SWBNT, one has to be relying on ab initio (or first-principles) simulation of SWBNTs and the relevant 2D boron sheet studies.

4.2.1 The Single-Layer Two-Dimensional Boron Sheets

A priori, the number of possible configurations for BNTs is enormous depending on their chirality, diameter, and number of tubular walls for the system, analogous to CNTs [46]. For a single-walled carbon nanotube (SWCNT), the constituent tubular wall is always referred to as a single-layer 2D graphene sheet. However, for an SWBNT, the constituent tubular wall can be more complicated because it is possible that not only one but multiple sheet structures are thermodynamically feasible, analogous to polymorphism of elemental boron solid in the crystalline phase. Thus by excluding the possible dominant effects among the wall-to-wall interactions along the tubular axis [31], to model an SWBNT, one has to consider different structural motif of an SWBNT from each of the different structural classes of single-layer 2D boron sheets [20, 21, 47].

Assuming that the formation of SWBNTs can be analogous to SWCNTs that are formed only under kinetically constrained conditions, one can conjecture that "one of the main difficulties in synthesizing BNTs appears to be the instability of a two-dimension graphene-like boron sheet." From its basic electron-deficient feature, it is reasonable to assume that a 2D boron sheet is generally a frustrated system that does not have enough electrons to fill all electronic orbitals in a chemical bonding that is based on pure sp^2 hybridization, and therefore, it is highly probable it consequently does not exhibit some clear preference for a simple structural motif that is similar to graphene of carbon. Hence, from an energetically perspective, there is no driving force toward some well-defined structure, and this might explain the difficulty in the synthesis of SWBNTs or single-layer boron sheets.

Thus while waiting for the experimental evidence, several ab initio or first-principles theoretical studies on 2D boron sheets have been proposed in recent years [18, 20, 21, 23, 47]. Since the elemental boron bulk solids have neither a purely covalent nor a purely metallic character, one can argue that the delocalized 3-c bonds and the electron-deficient features of boron atoms should be energetically more competitive and stable than homogeneous bonding character with only the sp^2 hybridization, as found in the carbon graphitic system. Hence, the competing roles played by the 3-c and 2-c bonding in determining the structural motifs of the 2D boron sheets were investigated recently [20].

Among the possible configurations that have been studied, it was found that there are at least three different structural classes of 2D boron sheets that are thermodynamic feasible, that is, the α-boron sheet, the buckled triangular sheet, and the distorted hexagonal sheet (Fig. 4.3) [20, 47]. From the α-B_{12} tetragonal boron bulk solid–derived 2D boron sheet, the system that consists of a B_{12} icosahedron in reduced dimension is found to be energetically slightly less favorable due to insufficient 2-c and 3-c intericosahedral bonds relative to the 3D bulk crystalline solid [20]. Among these theoretical predicted candidates, the α-boron sheet is found to be the most stable. The structure is representing a combination of both 3-c and 2-c bonding that composed of both hexagonal (2-c) and triangular (3-c) motifs (Fig. 4.2) with cohesive energy found to be about ~93% of the α-rhombohedral (α-B_{12}) boron solid [20]. The extra stability of this structural motif can be explained via the electronic structure analysis of the system proposed by Tang et al. [21].

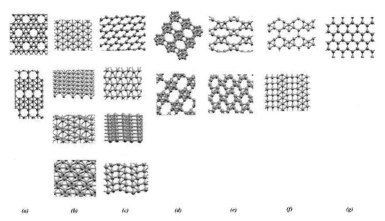

Figure 4.2 Two-dimensional boron sheets that consist of various structural motifs: (a) α- and β-sheet (i.e., a hybrid of triangular and hexagonal structure units), (b) flat and buckled triangular sheet, (c) distorted hexagonal (flat and buckled), (d) icosahedral unit–based boron sheet, (e) low symmetry, (f) mixed structural units (e.g., triangular, square, hexagonal, etc.), and (g) hexagonal graphene-like boron sheet.

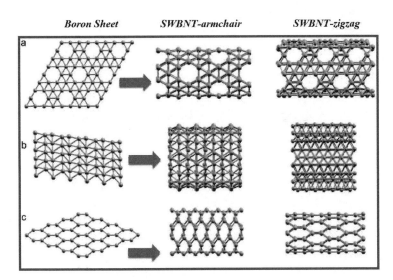

Figure 4.3 Atomic structure of boron sheets (first column) and armchair and zigzag BNTs (second and third columns, respectively). (a) Structure motif derived from the α-boron sheet, (b) the buckled triangular sheet, and (c) the distorted hexagonal boron sheet.

According to Tang et al. [21], balance interplay in an α-boron sheet from the competition between 2-c and 3-c bonding is necessary to preserve the bonding stability of this boron sheet. In an α-boron sheet, the hexagonal and triangular structural motifs have distinct electronic properties. The sp^2-state of hexagonal structural motif is partially unoccupied, therefore is electron-deficient and prone to accepting electrons, whereas the flat triangular structural motif is found to has a surplus of electrons in antibonding states. From a doping perspective, the 3-c flat triangular regions should act as donors, while the 2-c hexagonal regions should act as acceptors. Therefore, the stability of these 2D boron sheets depend strongly on the ratio of hexagons to triangles, which can be described by a hexagon hole density, η [21]. Under this model, η changes from 0 to 1/3 as the boron sheet alters from the triangular sheet T(0) to the ideal hexagonal graphene-like boron sheet H(1/3). Above all, the most stable boron sheet structure based on this structural motif is found to be an α-boron sheet or A(1/9), as shown in Fig. 4.2 [31], which occurs at η = 1/9. Thus this suggested that if the system is able to turn into a mixture of these two phases in the right proportion, it should benefit from the added stability of both subsystems.

Following the α-boron sheet, the next stable configuration is the buckled triangular sheet (Fig. 4.2). It was obtained from the geometry relaxation of a flat triangular sheet through a buckling along the perpendicular direction of the sheet. Subsequently, the buckling mixes in-plane and out-of-plane electronic states and can be thought of as a symmetry reducing distortion that enhances binding. Intuitively, the buckling of boron atoms is a response of the sheet to the internal stress imposed by the arrangement of the atoms in a perfect triangular 2D lattice. For the flat triangular 2D lattice, the delocalized 3-c bonds dominant structural motif is found to be less stable. According to a recent density functional theory (DFT) calculation [47], the atomic lattice of this structure is found to be dynamically unstable with imaginary phonon. In particular, all of its transverse out-of-plane and ZA bending modes are not stable with imaginary ω_c ranging from ~261 to 193 cm^{-1}. It therefore appears that the 3-c bonds present in the perfect triangular lattice are not strong enough to bind the boron atoms in a 2D lattice. In terms of electronic bonding characters, for a perfect triangular boron sheet, both the high atomic coordination and the electronic-deficient character of boron yield a nearly homogenous electron density distribution in the 2D lattice. Its Fermi level is high enough to force some of the electrons to occupy antibonding molecular orbitals, which, in turn, induce a destabilizing effect in the 2D lattice making it to be highly chemically reactive [21]. Thus through a buckling on the triangular boron sheet, it can stabilize both lattice and chemical bonding of this structural motif.

As proposed by Tang et al. [21], a perfect hexagonal lattice of boron in 2D sheet is not stable due to its partially occupied sp^2 state. The instability of this hexagonal boron sheet has been confirmed by a recent DFT phonon study [47]. However, by breaking the D_{6h} symmetry of the hexagonal lattice to a lower D_{2h} symmetry of the orthorhombic lattice, the reduced symmetry can make the sheet configuration to be relatively stable ($\sim$90% α-rhombohedral [α-B$_{12}$] boron solid cohesive energy) and subsequently yield the distorted hexagonal boron sheet [20, 47]. In terms of local geometry, this boron sheet is comprised of a network of triangle-square-triangle units in the lattice. Similar to the α-boron sheet and the buckled triangular sheet configurations, the chemical bonding in the distorted hexagonal sheet can also be characterized by both 2-c and 3-c bonds, reminiscent of the electron-deficient features of boron atoms in bulk solid.

According to Lau et al. [47], the ab initio DFT results suggest that the nature of the chemical bonding, rather than thermal effects, appears to be the prime factor in determining the stability of atomic monolayers of boron. Despite the fact that they (i.e., α-, buckled triangular, and distorted hexagonal boron sheet) are found to be thermodynamically stable, these 2D phases of elemental boron remain thermodynamically less stable than the 2D graphene of carbon. It is important to point out that thermodynamic stability of these boron sheets relative to boron crystalline bulk solid is substantially less stable than graphene relative to carbon bulk solid, that is, graphite. For a monolayer graphene, its cohesive energy is $\sim$98% of graphite, whereas for these monolayer boron sheets, they are found to be merely $\sim$90%–93% as stable as the commonly known boron bulk crystal, α-B$_{12}$ rhombohedral solid [20, 47]. Unlike the carbon, the stability of graphene sheets are mostly sp^2 bond driven, and only weakly bind by the interlayer long-range van der

Waals force in graphite for carbon bulk solid; however, this unique feature is not found for the boron atoms. Thus to fill up the missing energy gap between these 2D boron sheets and boron bulk solid, the strong interlayer interaction between the boron sheets is thus expected [31]. Similarly, the strong bonding among wall-to-wall interactions for the multiwalled BNTs [18, 31, 33] and the strong intertubular bonding interactions within the crystalline bundles of SWBNTs [28] are therefore predicted.

As anticipated from Fig. 4.2, the number of possible atomic configurations for an SWBNT can be enormous with respect to different structural motifs of boron sheets. However, to keep our discussion as vivid as possible, we shall only focus on our discussion on the energetically most stable α-boron sheet–derived SWBNT for the sake of simplicity in the description. Specifically, we will focus on their energetic stability, electronic properties, and elastic mechanical properties as predicted by DFT from literatures. For the SWBNTs that are derived from the buckled triangular boron sheet and the distorted hexagonal boron sheet–derived BNTs, the details can be found from the relevant papers [18, 19, 23, 28, 33].

4.3 Basic Properties of an SWBNT Derived from an α-Boron Sheet

4.3.1 Basic Geometry Construction

With a single-layer 2D α-boron sheet as a precursor, the SWBNTs rolled into a 1D hollow cylindrical shape with axial symmetry and in general exhibited a spiral conformation that called *chirality*. According to the diameter of an SWBNT, it defines the curvature of a nanotube, and chirality of the nanotube is usually defined by the chiral vector along the rolled direction (Fig. 4.4) analogous to an SWCNT. For an SWBNT that is derived from an α-boron sheet, as suggested by Yang et al. [29], two kinds of vectors can be used to determine the BNTs: the primary vectors (a) of the hexagonal lattice used for CNTs and the primary vectors of boron sheet lattice (b), as shown in Fig. 4.4. For the hexagonal lattice, $\mathbf{a}_1 = a\mathbf{x}$, $\mathbf{a}_2 = a(1/2\mathbf{x} + \sqrt{3}/2\mathbf{y})$, $\mathbf{a} = \sqrt{3}\, l_{\text{B-B}}$. For the α-boron sheet lattice, $\mathbf{b}_1 = b(3/2\mathbf{x} + \sqrt{3}/2\mathbf{y})$, $\mathbf{b}_2 = b(3/2\mathbf{x} - \sqrt{3}/2\mathbf{y})$, $b = 3\, l_{\text{B-B}}$. To satisfy the definition of the chiral vector, $\mathbf{C}_h = p\mathbf{b}_1 + q\mathbf{b}_2 = n\mathbf{a}_1 + m\mathbf{a}_2$, the chiral vector $\mathbf{C}_h$ (or equivalent to (p, q)) of the boron sheet lattice corresponds to the vector (n, m) of the hexagonal lattice with $p = (n + 2m)/3$ and $q = (n - m)/3$. Thus, the $(p, 0)$ BNT corresponds to the (n, n) with $(n = p)$ CNTs and the (p, p) BNT corresponds to the $(n, 0)$ with $(n = 3p)$ CNTs. The lattice constants are enlarged since the unit cell of the $(p, 0)$ BNTs contains three unit cells of the (n, n) with $(n = p)$ CNTs. To name an SWBNT, we follow the established standard for an SWCNT [46]; the $(p, 0)$ BNT is thus defined as an *armchair* SWBNT (i.e., α-SWBNT-arm), whereas for the *(p, p)* BNTs, it is thus defined as a *zigzag* BNT (i.e., α-SWBNT-zz). For the *(p, q)* BNTs, it is thus defined as a *chiral* SWBNT (i.e., α-SWBNT-chi). Coth zigzag and armchair SWBNTs are thus classified as *achiral*, whose image has an identical structure to the original one [46].

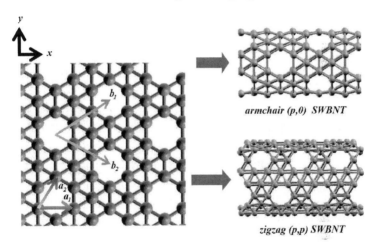

armchair (p,0) SWBNT

zigzag (p,p) SWBNT

Figure 4.4   The basic lattice vectors of an α-boron sheet ($\mathbf{b}_1, \mathbf{b}_2$) compared to graphene-like boron sheet lattice vectors ($\mathbf{a}_1, \mathbf{a}_2$), and the corresponding armchair and zigzag SWBNT follows the definition from Ref. [29].

4.3.2   Energetic Stability

To evaluate the basic properties of SWBNTs from simulation, we now first turn to the energetic stability of an SWBNT that can be determined accurately on the basis of DFT. In this case, the energetic stability of an SWBNT can be predicted on the basis of the both cohesive energy, E_{coh}, and the curvature energy, E_{curv} (or the strain energy), from DFT calculation. In this case, the cohesive energy is in principle equivalent to the binding energy or atomization energy) of a system that defined as $E_{coh} = -E_{tot}/N + E_{atom}$, where E_{tot} and E_{atom} are the ground-state energies of the whole system and an isolated boron atom, respectively, and N is the number of boron atoms in the system. From this definition it follows that the positive values of E_{coh} correspond to a bound (or energetically stable) tubular structures. To a first approximation, the chemical stability of a certain structure can be judged by E_{coh}. The ground-state energies in general can be obtained with DFT and the geometries of all the considered structures have to be fully optimized. As shown in Fig. 4.5, the SWBNTs that derived from an α-boron sheet are generally less stable than their corresponding α-boron sheet (i.e., ~93%–94% of α-B_{12} rhombohedral bulk solid [20, 29]) but are more stable than the B_{80} fullerene structure due to less strain in the curvature [29].

The curvature energy of an SWBNT, E_{curv}, is defined as $E_{curv} = E_{tube} - E_{sheet}$, where E_{tube} is the energy per atom of the nanotube and E_{sheet} is the energy per atom of the corresponding boron sheet structure (i.e., the α-boron sheet). Therefore, E_{curv} can be comprehended as the energy cost of rolling up the sheet into a specific (p, q) nanotubular structure. The curvature effects decrease with increasing diameter, and therefore E_{curv} tends to be zero for SWBNTs with large diameters. Under the elastic theory, the curvature energies of both armchair $(p, 0)$ and zigzag (p, p) types of SWBNTs in general should lie essentially on a line and follow the $E_{curv} = C/D^2$ relation, where C is a fitting constant and D is the diameter of the corresponding nanotube. Regardless of different chirality of SWBNTs that are derived from an α-boron sheet, the $E_{curv} \approx 1/D^2$ relation generally can be found for nanotubes with diameter $D > 5$ Å. If different classes of SWBNTs are found to lie almost on the same curve following their E_{curv}, then this suggests that the curvature energy of the nanotubes should in principle not vary strongly based on different chiralities for the SWBNTs. Interestingly this simple observation is only found to be true for the achiral (i.e., $(p, 0)$ and (p, p)) and chiral (i.e., (p, q)) SWBNTs that are derived from the α-boron sheet [29, 31, 33] but not for the SWBNTs that derived from the buckled triangular and distorted hexagonal boron sheets. For these two structural motif–based SWBNTs, it was found that the E_{curv} depends not only on the tube's diameter but also on its chirality [18, 19, 23, 33].

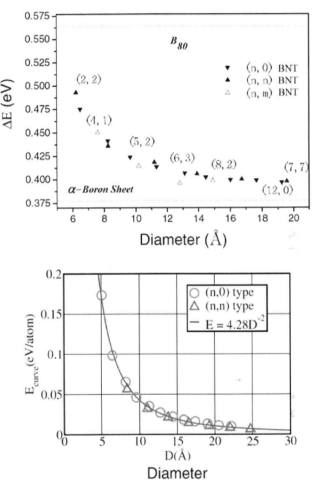

Figure 4.5 (Top) The relative formation energies of SWBNTs vs. the diameters. The formation energy per atom is measured with respect to the α-B_{12} rhombohedral solid. The relative formation energies of the B_{80} fullerene and the α-boron sheet are plotted for comparisons, as adopted from Ref. [29]. (Bottom) The curvature energies E_{curv} in eV/atom of SWBNTs derived from an α-boron sheet vs. diameter D (in Å) together with a numerical fitted line in $1/D^2$ from Ref. [31].

As in the case for SWCNTs, we expect from elastic theory that E_{curv} should have the following simple dependence on $\sim 1/D^2$ for a large enough diameter, D, which follows $E_{curv} = C/D^2$, where C is a constant. By fitting the DFT values of E_{curv}, one can obtain the C value. As found by different authors [30, 31], the C values are reported to be within a range of $\sim$3.64–4.28 eVÅ^2/atom for both zigzag and armchair SWBNTs. For comparison, SWCNTs are reported to have $C \approx 8.56$ eV Å^2/atom [48, 49]. Therefore, for a given diameter, one can claim that it might be easier to roll an α-boron sheet than to curve graphene to create SWCNTs. However, it remains to be determined by experiments to prove whether such a prediction is feasible. In addition, it is noteworthy to point out that boron's well-known polymorphism that observed in its solid phase also appears to be true for small-diameter SWBNTs (i.e., $D < 5$ Å) [33]. In this regime, the nanotubes that derived from the three stable boron sheets (i.e., α-, buckled triangular, and distorted hexagonal boron sheets shown in Figs. 4.2 and 4.3) are very close in cohesive energy and competing to each other, and thus for small-radii SWBNTs, their relative stability is strongly dependent on their diameters, chirality, and structural motifs [33].

4.3.3 Electronic Properties

For an SWCNT, it is generally known that its electronic properties can be metallic or semiconductor, depending on the nanotube chirality [46]. In contrast to carbon graphene, which is semimetallic in nature, the thermodynamically most stable α-boron sheet is found to be metallic (Fig. 4.6) with finite electronic density of states at the Fermi level from the out-of-plane π-manifold [21]. This finding suggests that in contrast to SWCNTs, a pristine SWBNT should always be metallic according to the zone-folding technique, irrespective to its chirality [46]. However, as the diameter of SWBNTs is decreased, the differences between the prediction of zone folding and DFT calculations become substantial. As shown in Fig. 4.6, the small-diameter (3,3) SWBNT (i.e., $D \approx 8$ Å) is predicted to be semiconducting with a small band gap, whereas it is found to be metallic according to zone-folding techniques. According to the recent DFT results, the small-diameter pristine

SWBNT (i.e., $D < 15$ Å) that is built from the α-boron sheet can be a semiconductor, with band gaps staying around a few tenths of eVs, due to high curvature and surface buckling [29, 31] (Fig. 4.6). According to DFT prediction, a thermodynamic favorable metallic-to-semiconducting transition generally can be found in small-diameter SWBNTs (e.g., (3,0) SWBNTs with $D \approx 5$ Å in Fig. 4.6) due to the surface buckling attributed to rehybridization in the σ-π manifold that is similar to small-radii SWCNTs.

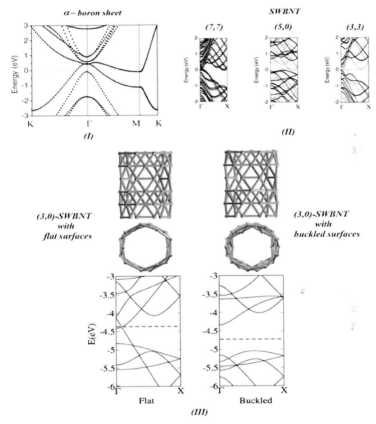

Figure 4.6   (I) The band structure of a metallic-like α-boron sheet; (II) the band structures of (7,7), (5,0), and (3,3) SWBNTs obtained from zone-folding (red lines) and DFT (black dots) calculations with Fermi level set to be zero [29]; and (III) the band structures of (3,0) SWBNTs (top and side view) with flat and buckled surfaces obtained from DFT calculations. The red dash lines show the Fermi level [31].

According to DFT calculations, for the nanotube diameters larger than $\sim$15 Å, most of the SWBNTs are metals [29, 31] (Fig. 4.7). This finding is qualitative consistent with the reported BNTs from experiment, that is, the metallic BNTs with larger diameters, that is, $\sim$10–40 nm [34, 35]. According to DFT predictions [29], when the diameters of the SWBNTs are smaller than 15 Å, the $(p, 0)$ boron single-walled nanotubes are semiconductors and the band gap of the $(p, 0)$ nanotubes with p = 4–9 decreases from $\sim$0.78 eV to 0.27 eV as the diameter increases. The (p, p) nanotubes with p = 3–6 are semiconductors with smaller band gaps from $\sim$0.21 eV to 0.11 eV. This trend generally holds true for both the local density approximation (LDA) and the generalized gradient approximation (GGA) in DFT calculations [29, 31]. As anticipated at Fig. 4.7, for the nanotubes with similar diameters, the band gap decreases as the chiral angle increases because the $(p, 0)$ nanotubes have largest gaps, while the (p, p) nanotubes have smallest gaps. As is shown in Fig. 4.6, there are electronic bands similar to the π-band in the graphene near the Fermi level, which is degenerated at the Γ-point about 0.5 eV above the Fermi level. From Fig. 4.6, there are several intersections between the Fermi level and electronic bands for the 2D α-boron sheet, in contrast to carbon graphene. Thus, the zone-folding features for SWBNTs would be more complex compared to the CNTs [46].

For large-diameter SWBNTs, the band structures obtained by zone folding are generally in good agreement with the ones obtained from DFT calculations, as shown for the metallic (7, 7) nanotube in Fig. 4.6. Due to the fact that the σ-π manifold rehybridization is minimal in flat and small curvature of a boron sheet, the band structures obtained from zone folding and DFT calculations are generally found to be equivalent for large-radii BNTs. As anticipated from Fig. 4.6, for the smaller-diameter nanotubes (e.g., $D <$ 15 Å), the predicted band structures obtained by zone folding generally break down and one has to rely on DFT calculations. According to zone folding, the small-diameter nanotubes are metals, while the DFT calculations predict that the nanotubes are generally semiconductors. From Fig. 4.6, the (5, 0) SWBNT is a semiconductor with an indirect gap according to DFT, but it is metallic on the basis of zone-folding prediction. The electronic band structure of the (3, 3) SWBNT shown in Fig. 4.6 is another example of semiconducting nanotubes with a direct band gap that zone folding failed in making the correct prediction of. In this case, the failure of the zone folding can attributed to substantial quantum confinement effect at small-diameter BNTs. In addition, significant local structure deformation and surface buckling that leads to symmetry breaking can generally be found in small-radii nanotubes, for example, the (3, 0) SWBNT shown in Fig. 4.6. As reported in the recent DFT calculations, the nanotubular surfaces become buckled under the large curvature necessitated by the small diameter. The buckling has been attributed to rehybridization in the σ-π manifold of the α-boron sheet, and subsequently leads to opening up of the band gap in electronic properties [29–31].

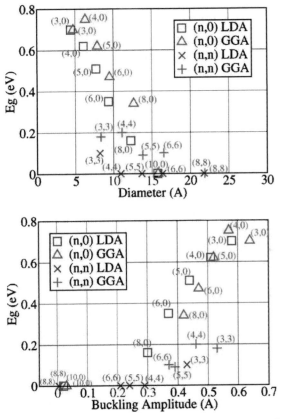

Figure 4.7 (Top) Band gap, E_g (in eV), vs. tube diameter (in Å) and the degree of buckling amplitude (bottom) (in Å) for $(n, 0)$ and (n, n) SWBNTs obtained from DFT LDA and GGA calculations from Ref. [31].

As pointed out by Tang et al. [31], the lowest-energy α-boron sheet (i.e., A(1/9) boron sheet as mentioned in Ref. [31]) prefers to stay flat when it is stress free but will buckle under compression. When it buckles, the two boron atoms in each unit cell that are in the triangular regions will move out of the sheet plane, with one going up and the other going down [31]. When the corresponding boron sheet is folded to form SWBNTs, a similar buckling pattern is observed: the two atoms in the triangular regions become inequivalent, with one moving radially inward and the other moving radially outward. Due to this surface buckling, the small-diameter SWBNTs become semiconducting (e.g., the (3, 0) tubes in Fig. 4.6). In general, this buckling-induced semiconducting behavior is only relevant for small-diameter SWBNTs, as shown in Fig. 4.6. The band gap vanishes once the BNT diameters are larger than $\sim$15 Å. As shown in Fig. 4.7, the qualitative trends of LDA and GGA prediction is generally similar; however, the predicted band gaps are systematically different by $\sim$0.1–0.2 eV for the same nanotube index in chirality [31]. The LDA predicted band gaps are generally smaller, and GGA generally predicts larger surface buckling than LDA (Table 4.1 and Fig. 4.7). According to Tang et al. [31], for all the SWBNTs, the surface buckling is always accompanied by a decrease in kinetic energy E_{kin} and increases in E_{xc} (exchange correlation energy) and E_{es} (classical electrostatic energy that includes electron–electron, electron–ion, and ion–ion interaction), the same as for the 2D α-boron sheets. Therefore it is reasonable to argue that the surface buckling in SWBNTs for a small diameter might also be driven by lowering of kinetic energy of the electron gas density that constitutes the total electronic energy (E_{tot}) of an SWBNT, that is, $E_{tot} = E_{kin} + E_{xc} + E_{es}$.

As given in Table 4.1, the energy differences per atom δE (in meV/atom between the flat- and buckled-surface SWBNT) are extremely small and decrease to zero rapidly with increasing nanotube diameter. It is found that the δE is at most $\sim$15 meV/atom for the smallest-diameter nanotube with $D \approx 4.7$ Å (i.e., for (3, 0) SWBNT in Table 4.1). It is noteworthy to point out that GGA systematically favors buckled surfaces compared to LDA and predicts larger energy gains from buckling. Given the extremely small magnitude of these energy differences, thus it is not clear whether LDA or GGA is accurate enough to capture them correctly. In fact, a recent theoretical work has investigated the geometry of a finite (5, 0) SWBNT segment using the presumably more accurate second-order Møller–Plesset perturbation (MP2) method and concluded that no buckling exists on the surface of this finite nanotube because the buckled configuration has a higher energy [50], and is in contrast to the recent reported DFT results that are all based on LDA and GGA findings [29–31]. Hence, it is important to note that more caution might be required regarding the LDA/GGA predictions for such small energy differences and further theoretical studies are definitely required to determine the correct ground-state geometries and electronic properties of these small-radii SWBNTs that derived from the α-boron sheets.

Table 4.1 Energetic, structural, and electronic data for $(n, 0)$ and (n, n) SWBNTs derived from an α-boron sheet (i.e., A(1/9) sheet adapted from Ref. [31]) on the basis of LDA and GGA calculations. The table shows total energy difference δE (in meV/atom) between flat- and buckled-surfaced cases of an SWBNT, (average) nanotube diameter D (in Å), buckling amplitude A_{buckle} (in Å), and the nanotube band gap E_{gap} (in eV)

	LDA				GGA			
SWBNT	δE	D	A_{buckle}	E_{gap}	δE	D	A_{buckle}	E_{gap}
(3, 3)	2.71	8.20	0.43	0.10	6.57	8.3	0.53	0.18
(4, 4)	1.30	10.9	0.29	0	4.19	11.1	0.46	0.20
(5, 5)	0.68	13.7	0.24	0	2.73	13.8	0.39	0.09
(6, 6)	0.35	16.4	0.21	0	1.95	16.5	0.37	0.10
(8, 8)	0.32	21.8	0.01	0	1.39	21.9	0.01	0
(3, 0)	9.81	4.40	0.58	0.70	14.85	4.73	0.64	0.70
(4, 0)	7.16	6.06	0.51	0.62	11.38	6.31	0.57	0.75
(5, 0)	5.12	7.70	0.44	0.51	8.86	7.89	0.52	0.62
(6, 0)	3.13	9.33	0.37	0.35	6.32	9.47	0.47	0.47
(8, 0)	0.50	12.2	0.30	0.16	3.88	12.6	0.42	0.34
(10, 0)	0.03	15.7	0.02	0	2.29	15.8	0.03	0

Adapted from Ref. [31] with the permission of the American Physical Society (APS).

4.3.4 Elastic Properties

In this section, we shall turn our discussion to another basic yet interesting physical property that is studied the most on a 1D nanotubular structure, that is, the elastic behavior that determines the basic mechanical properties of nanotubes. To be practically realizable and possibly useful in device applications, the 1D nanotube should be able to sustain some mechanical deformations and be flexible in elastic properties in response to external forces and pressure. In graphite and in CNTs, there are three kinds of forces between carbon atoms that determine their characteristic elastic properties, that is, strong σ-bonding, π-bonding between the intralayer C=C bonds, and the weak interlayer van der Waals interaction. For graphite and CNTs, the strong sp^2 covalent bonds dominate the σ-bond skeleton of the honeycomb lattice, and thus the strength in the direction of the CNT axis is maximal, and this subsequently yield a high Young's modulus, $Y \approx C_{11}$ of graphite (i.e., ~ 1000 GPa) [46] due to their strong localized σ-bond. In spite of this unique feature, the CNT is not a rigid tubular object either. The cross section of a CNT is highly flexible and easily deformed when an external force is applied normal to the nanotube axis. When the diameter of an SWCNT increases, the nanotube is less stable in the direction perpendicular to the surface; therefore a flattened or bent nanotube usually can be observed [46].

Relative to comparatively more well-studied carbon graphitic materials and CNTs, here it is noteworthy to address that the intrinsic mechanical properties of BNTs remain elusive. So far, a systematic experimental study on the mechanical properties of BNTs cannot be found, and only very limited theoretical studies of mechanical properties of BNTs can be obtained in the literature [18, 28, 30]. To get a glimpse of the basic mechanical properties of SWBNTs on the basis of the structural motif of an α-boron sheet, we can only refer to a recent ab initio study [30]. On the basis of the DFT plane-wave calculations under the GGA model, the calculated stiffness C and Poisson ratio (v) of SWBNTs are shown in Table 4.2.

On the basis of the assumption of an isotropic and homogeneous behavior of materials within the framework of conventional linear elastic solid model, several important elastic constants of SWBNTs, such as Y (Young's modulus or modulus of elasticity), v (Poisson ratio), and B (bulk modulus or modulus of compression), can be predicted through a constitutive stress-strain relations within a set of dual independent elastic constants, that is, (Y, v). By varying the lattice constant of an SWBNT, the v is defined as a ratio of lateral strain and axial strain as follows: $v = -1/\varepsilon$ $(R - R_{eq}/R_{eq})$, where ε is the axial strain, R_{eq} is the equilibrium nanotube radius, and R is the nanotube radius at strain ε. As shown in Table 4.2, we find in all cases the Poisson ratio is positive (i.e., $v \approx 0.15$–0.25), namely, an elongation of the SWBNT reduces its diameter. The high Poisson ratio of the SWBNT implies that the diameter of the tube may change significantly as the tube is strained longitudinally. These values are slightly smaller than SWCNTs (i.e., $v \approx 0.23$–0.30) [51–53], and can be very much different from the SWBNTs that derived from triangular and distorted hexagonal boron sheets obtained from DFT values (i.e., ~0.10–0.50) [18, 54], which might attributed to the distinct stress and strain response of boron chemical bonding of different boron sheets.

Table 4.2   DFT values of the calculated SWBNT stiffness (C), diameter, and Poisson ratio (v)

SWBNT	Diameter (Å)	C (N/m)	v
(5, 5)	8.13	209.4	0.18
(9, 0)	8.63	206.7	0.21
(6, 6)	9.93	202.1	0.26
(12, 0)	11.31	204.6	0.21
(7, 7)	11.37	215.2	0.20
(8, 8)	13.13	214.0	0.21
(18, 0)	16.50	217.5	0.15

Adopted from Ref. [30].

Regarding the other important mechanical characteristics of BNTs, the modulus of elasticity or Young's modulus, Y, in its conventional definition is $Y = 1/V_0$ $(\partial^2 E/\partial \varepsilon^2)_{\varepsilon=0}$, where V_0 is the equilibrium volume and E is the total energy of the system from ab initio or first-principles calculations of the system. However, in the case study of Singh et al. [30], the uniaxial stiffness constant C is computed (Table 4.2) instead of Young's modulus, Y. In this case, the stiffness constant C of an SWBNT is a measure of the resistance due to elastic deformation. For a single degree of freedom (e.g., strain or compression of an SWBNT), it is defined as $C = F/\delta = 1/a\ (\partial^2 E/\partial \varepsilon^2)_{\varepsilon=0}$, where F is the force applied on an SWBNT, δ is the variation in the lattice constant, and a is the area per atom of the SWBNT. By computing the total energy E per atom as a function of elongation ε under uniaxial tension and within the harmonic approximation, for the total energy per atom (E) as a function of elongation strain ε with the relation E vs. ε, the C value of an SWBNT can be obtained by fitted to a parabola $E \approx 1/2C\varepsilon^2$ [30]. From Table 4.2, the stiffness of SWBNT shows a small variation (<10%) with an approximate value $C \approx 210$ N/m, regardless of the chirality and diameters. However, to compare it with a known mechanical property (i.e., elastic modulus) of SWCNTs that typically fall in the range of $Y \approx 0.78$–1.10 TPa [51–55], a relation between C and Y has to be known. In this case, for the special case of unconstrained uniaxial tension or compression, the elastic modulus, Y, can be thought as a measure of the stiffness, C, of a material. Thus under this model, the uniaxial stiffness, C, can be related to the elastic modulus, Y, of a nanotube as follows: $C = AY/L$, where A is the cross-sectional area and L is the length of the elemental unit of the nanotube. Thus by knowing the A, L, and C values of an SWBNT computed from DFT, the Young's modulus of an SWBNT that is derived from an α-boron sheet can be predicted. In this aspect, the effect on Young's modulus corresponding to the surface buckling of small-radii SWBNTs and the possibly strong interaction among the intertubular walls of multiwalled BNTs remain to be studied. To determine these elastic moduli of BNTs, accurate characterization and prediction in the variation of surface area and cross-sectional area of nanotubes are therefore critically important. Thus by considering the other energetic competing structural motifs of BNTs, this perhaps will open up a broader range of interesting research problems to the fundamental studies on the mechanical responses of elemental boron structures in the future.

4.4 Conclusion and Outlook

In short, there is an intriguing and potentially significant difference between carbon and boron due to their fundamental difference in valence electrons of individual atomic orbitals. From its basic electron-deficient feature, it is reasonable to assume that a 2D boron sheet is generally a frustrated system that does not have enough electrons to fill all electronic orbitals in a chemical bonding based on pure sp^2 hybridization, and therefore, it is highly probable it consequently does not exhibit some clear preference for a simple structural motif that is similar to graphene of carbon. Hence, from an energetically perspective, there is no driving force toward some well-defined structure, and this might explain the difficulty in the synthesis of SWBNTs or single-layer boron sheets. From the fact that no elemental boron but only compounds containing boron can be found on earth, together with the nature of their electron-deficient chemical bonds, might indicate that these boron structures are chemically reactive and will therefore not easily be found under ambient growth conditions. Thus, future study on the catalyst promote growth environment seems to be necessary.

However, on the basis of accurate DFT calculations, the predicted stability of the boron sheet configurations suggests, in principle, the feasibility of synthesis of BNTs. But it is important to point out that the thermodynamically most stable boron sheet (i.e., α-boron sheet) relative to boron crystalline bulk solid remains substantially less stable than graphene relative to carbon bulk solid, that is, graphite. For a monolayer graphene, its cohesive energy is ~98% of graphite, whereas for the α-boron sheet, it is found to be merely ~93%, as stable as the commonly known boron bulk crystal, α-B_{12} rhombohedral solid. Thus to fill up the missing energy gap between this 2D boron sheet (and the derived SWBNTs) and boron bulk solid, the strong interlayer interaction between the boron sheets and the derived BNTs are thus highly probable, and this speculation has recently been confirmed on the basis of DFT results on double-walled α-boron sheets and their double-walled BNTs [31]. From DFT, the formation of multiwalled BNTs seems favored over the SWBNTs due to their comparatively lower in cohesive energy (i.e., ~90% of thermodynamically stable α-B_{12} boron solid) for the latter. For the multiwalled BNTs with three or more layers, the convergence of their cohesive energies relative to α-B_{12} boron solid, however, remains to be studied. As a consequence, we may speculate that with proper choices of constituent sheets making up the walls, interwall covalent bonds will form and stabilize the overall configuration of these new nanostructures. In principle, one can imagine this process will continue, leading to highly stable BNTs with very many walls approaching experimental observation in the large-diameter regime, that is, 10–40 nm. However, the available experimental transmission electron microscopy shows that the actual fabricated BNTs are hollow in the center. If the ground-state structures of BNTs have many walls, this will raise the intriguing question, At what point kinetic limitations during the growth limit the available structures sampled during the fabrication? Therefore, we believe that a better understanding of the experimental growth mechanism and its implications on the fabricated structures is necessary for proper modeling and understanding of the properties of BNTs.

From the energetically most stable 2D boron sheet (i.e., α-boron sheet), the SWBNTs seems to be feasible and exhibit various interesting physical properties similar to the well-known CNTs. From molecular electron transport properties to electron transmission work function to metal-doped BNTs as high-hydrogen-storage materials, etc., their potential applications are currently being intensely studied in theory and often initiate interesting discussion. However, not all these studies advanced to explain them stand up to critical examination, unless the BNTs' growth mechanism is properly understood. It should also be stressed that a critical examination of the fundamentals on BNTs and boron sheets on theory and experiments would often be desirable, and much remains to be done. Above all, the solid-state chemistry and polymorphism of the element boron is full of booby traps at all levels from macroscopic bulks to nanostructures, none of which detract from its charms.

References

1. Iijima S., *Nature (London)*, **354**, 56 (1991).
2. Geim A. K., and Novoselov K. S., *Nature Materials*, **6** (3), 183–191 (2007).
3. Kroto H. W., Heath J. R., O'Brien S. C. O., Curl R. F., and Smalley R. E., *Nature (London)*, **318**, 162–163 (1985).
4. Senger R.T, Dag S., and Ciraci S., *Physical Review Letters*, **93**, 196807 (2004).
5. Oshima Y., Onga A., and Takayanagi K., *Physical Review Letters*, **91**, 205503 (2003).

6. Li Y., Wang J., Deng Z., Wu Y., Sun X., Yu D., and Yang P., *Journal of the American Chemical Society*, **123**, 9904 (2005).

7. Yang X., and Ni J., *Physical Review B*, **72**, 195426 (2005).

8. Crescenzi M. De., Castrucci P., and Scarselli M., *Applied Physics Letters*, **86**, 231901 (2005).

9. Carroll D. L., Redlich, Ph., Blase X., Charlier J.-C., Curran S., Ajayan P. M., Roth S., and Rühle M., *Physical Review Letters*, **81**, 2332 (1998).

10. Miyamoto Y., Rubio A., Louie S. G., and Cohen M. L., *Physical Review B*, **50**, 18360 (1994).

11. Xiang H. J., Yang J., Hou J. G., and Zhu Q., *Physical Review B*, **68**, 035427 (2003).

12. Bengu E., and Marks L. D., *Physical Review Letters*, **86**, 2385 (2001).

13. Seifert G., Terrones H., Terrones M., Jungnickel G., and Frauenheim T., *Physical Review Letters*, **85**, 146 (2000).

14. Tenne R., Margulis L., Genut M., and Hodes G., *Nature (London)*, **360**, 444 (1992).

15. Otten C. J., Lourie O. R., Yu M., Cowley J. M., Dyer M. J., Ruoff R. S., and Buhro W. E., *Journal of the American Chemical Society*, **124**, 4564 (2002).

16. Xu T. T., Zheng J., Wu N., Nichollas A. W., Roth J. R., Dikin D. A., and Ruoff R. S., *Nano Letters*, **4**, 963 (2004).

17. Kirihara K., Wang Z., Kawaguchi K., Shimizu Y., Sasaki T., Koshizaki, N., Soga K., and Kimura K., *Applied Physics Letters*, **86**, 212101 (2005).

18. Evans M. H., Joannopoulos J. D., and Pantelides S. T., *Physical Review B*, **72**, 045434 (2005).

19. Lau K. C., Pati R., Pandey R., and Pineda A. C., *Chemical Physics Letters*, **418**, 549 (2006).

20. Lau K. C., and Pandey R., *The Journal of Physical Chemistry C*, **111**, 2906 (2007).

21. Tang H., and Ismail-Beigi S., *Physical Review Letters*, **99**, 115501 (2007).

22. Kunstmann J., and Quandt A., *Chemical Physics Letters*, **402**, 21 (2005).

23. Kunstmann J., and Quandt A., *Physical Review B*, **74**, 035413 (2006).

24. Boustani I., and Quandt A., *Europhysics Letters*, **39**, 527 (1997).

25. Gindulytu A., Lipscomb W. N., and Massa L., *Inorgcanic Chemistry*, **37**, 6544 (1998).

26. Boustani I., Quandt A., Hernandez E., and Rubio A., *The Journal of Chemical Physics,* **110**, 3176 (1999).

27. Lau K. C., Pandey R., Pati R., and Karna S.P, *Applied Physics Letters,* **88**, 212111 (2006).

28. Lau K. C., Orlando R., and Pandey R., *Journal of Physics: Condensed Matter,* **20**, 125202 (2008).

29. Yang X., and Jun N., *Physical Review B* (R), **77**, 041402 (2008).

30. Singh A. K., Sadrzadeh A., and Yakobson B. I., *Nano Letters,* **8**, 1314 (2008).

31. Tang H., and Ismail-Beigi S., *Physical Review B,* **82**, 115412 (2010).

32. Lau K. C., Orlando R., and Pandey R., *Journal of Physics: Condensed Matter,* **21**, 045304 (2009).

33. Bezugly V., Kunstmann J., Grundkotter-Stock B., Frauenheim T., Niehaus T., and Cuniberti G., *ACS Nano,* **5**, 4997 (2011).

34. Liu F., Shen C., Su Z., Ding X., Deng S., Chen J., Xu N., and Gao H., *Journal of Materials Chemistry,* **20**, 2197–2205 (2010).

35. Ciuparu D., Klie R. F., Zhu Y., and Pfefferle L., *The Journal of Physical Chemistry B,* **108**, 3967–3969 (2004).

36. Muetterties E. L. (Ed.) *The Chemistry of Boron and Its Compounds* (John Wiley, New York, 1967).

37. Muetterties E. L. (Ed.) *Boron Hydride Chemistry* (Academic, New York, 1975).

38. Pauling L., *Nature of Chemical Bond and the Structure of Molecules and Crystals,* 3rd. Ed., (Cornell University Press, Itacha, 1960).

39. De S., Willand A., Amsler M., Pochet P., Genovese L., and Goedecker S., *Physical Review Letters,* **106**, 225502 (2011).

40. Boustani I., *Physical Review B,* **55**, 16426 (1997).

41. Ozdogan C. O., Mukhopadhyay S., Hayami S., Guvenc Z. B., Pandey R., and Boustani I., *The Journal of Physical Chemistry C,* **114**, 4362 (2010).

42. Szwacki N. G., Sadrzadeh A., and Yakobson B. I., *Physical Review Letters,* **98**, 166804 (2007).

43. Parakhonskiy G., Dubrovinskaia N., Bykova E., Wirth R., and Dubrovinsky L., *Scientific Reports,* **1**, 96 (2011).

44. Fujimori M., Nakata T., Nakayama T., Nishibori E., Kimura K., Takata M., and Sakata M., *Physical Review Letters,* **82**, 4452 (1997).

45. Albert B., and Hillebrecht H., *Angewandte Chemie International Edition,* **48**, 8640 (2009).

46. Saito E. Dresselhaus G., and Dresselhaus M. S., *Physical Properties of Carbon Nanotubes* (Imperial College Press, London, UK, 2003).

47. Lau K. C., and Pandey R., *The Journal of Physical Chemistry C,* **112**, 10217 (2008).

48. Robertson D. H., Brenner D. W., and Mintmire J. W., *Physical Review B,* **45**, 12592 (1992).

49. Gülseren O., Yildirim T., and Ciraci S., *Physical Review B,* **65**, 153405 (2002).

50. Gonzalez Szwacki N., and Tymczak C. J., *Chemical Physics Letters,* **494**, 80 (2010).

51. Hernandez E., Goze C., Bernier P., and Rubio A., *Physical Review Letters,* **80**, 4052 (1998).

52. Lu J. P., *Physical Review Letters,* **79**, 1297 (1997).

53. Sanchez-Portal D., Artacho E., Soler J. M., Rubio A., and Ordejon P., *Physical Review B,* **59**, 12678 (1999).

54. Lau K. C., *First-Principles Study of Boron Nanostructures,* PhD thesis (2007).

55. Yu M. F., Files B. S., Arepalli S., and Ruoff R. S., *Physical Review Letters,* **84**, 5552 (2000).

Chapter 5

Boron Nanowires: Synthesis and Properties

Shobha Shukla
Department of Metallurgical Engineering and Materials Science,
Indian Institute of Technology Bombay, Mumbai 400076, India
sshukla@iitb.ac.in

5.1 Introduction

Boron is the only nonmetallic element with a band gap of 1.56 eV in group III of the periodic table, having properties similar to its neighbor carbon and diagonal relative silicon. It is electron deficient and possesses a vacant p-orbital. Its small size and high ionizing energy favors the formation of covalent bonds rather than metallic bonding. Bulk boron is found to exist into well-known crystalline α-rhombohedral, β-rhombohedral, and β-tetragonal form; however, it is also known to exist in different amorphous forms. Under specialized conditions α-tetragonal and γ-orthorhombic phases can also be synthesized. This trivalent element is characterized by its short covalent radius and has the tendency to form strong directional chemical bonds to produce molecular compounds. This makes boron very fascinating element

Handbook of Boron Nanostructures
Edited by Sumit Saxena
Copyright © 2016 Pan Stanford Publishing Pte. Ltd.
ISBN 978-981-4613-94-1 (Hardcover), 978-981-4613-95-8 (eBook)
www.panstanford.com

in the periodic table. Boron compounds are composed of quite regular icosahedral structural unit, as shown in Fig. 5.1, through a unique three-center two-electron bond, which provides unique physical and chemical properties to boron [1, 2]. At standard temperatures it is a poor electrical conductor but shows good conduction properties at high temperatures. It is the third-lightest element with a low density of 2.340 g/cm^3, a high melting point of 2300°C, a large Young's modulus of 380–400 GPa, high chemical stability, and hardness close to that of diamond. This makes boron and its related compounds promising for fundamental research [3–5] along with applications in thermoelectric power conversion devices, as a lightweight coating for space shuttles [6], neutron absorbent [7], nuclear engineering [5, 6], photoelectronics, and other futuristic nanodevices [8–14].

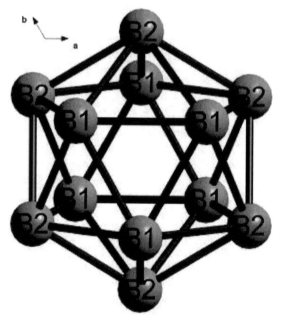

Figure 5.1   Boron icosahedron structural unit [2].

One-dimensional boron nanostructures have received steadily growing interest as a result of their potential fundamental characteristics and peculiar applications in areas such as materials science, physics, chemistry, bioscience, and industry as compared to their bulk counterparts. One-dimensional flexible nanomaterials that can retain excellent electrical performance even under large mechanical strain are very useful for the miniaturization of future nanoelectronic devices. Theoretical predictions have predict that among other 1D nanomaterials, boron is especially attractive due to its interesting electrical and mechanical properties [15, 16]. One-dimensional boron nanostructures and metal borides possess comparable mechanical strength, chemical stability, thermal stability, and electrical conductivities to those of carbon nanotubes and other related families of nanosystems [4, 14]. Figure 5.2 below shows boron in the form of nanowires [15], nanocones [9], and nanobelts [17].

Figure 5.2 (a) SEM image of crystalline boron nanowires [15], (b) TEM image of a boron nanocone [9], and (c) SEM image of a boron nanobelt [17].

Of all these 1D nanosystems, studies on boron nanowires have recently gained momentum and are being considered as core components for high temperature semiconducting nanodevices and other futuristic nanodevices.

5.2 Synthesis of Boron Nanowires

Various physical and chemical methods have been utilized to prepare 1D boron nanostructures such as nanowires [14, 15, 18], nanotubes [19], and nanobelts [17]. Three different techniques that have explicitly been utilized to synthesize boron nanowires are (1) chemical vapor deposition (CVD) [4, 15, 18, 20–24], (2) magnetron sputtering (MS) [1, 25–29], and (3) laser ablation (LA) [30–34]. The reported synthesis conditions for boron nanowires using these techniques is summarized in Table 5.1, Table 5.2, and Table 5.3 [5].

Table 5.1 Synthesis conditions of boron nanowires by magnetic sputtering [5]

Precursor	Carrier gas	T (°C)	Catalyst	Substrate	Diameter	Structure
B/B_2O_3 plate	Ar	500–800	Au film	Si (100)	10–20	Crystalline
B/B_2O_3 (40 wt.%)	Ar	800		Si wafer	40	Amorphous array
B target	Ar	800		Si wafer	40–60	Amorphous array
B/B_2O_3 target		800–900		Si wafer	20–40	Amorphous feather-like

Table 5.2 Synthesis conditions of boron nanowires using CVD techniques [5]

Precursor	Carrier gas	T (°C)	Catalyst	Substrate	Diameter	Structure
B, I_2, Si Powder		1000–1100	Au film	MgO	50–100	Amorphous
B_2H_6/Ar (5%)	Ar	1100	NiB	Alumina	20–200	Crystalline
Ar/H_2/B_2H_6	H	800	NCA	NCA	20–60	Tetragonal
B, I_2, Si powder		900–1200	Au film	Si (100)	40–200	Polycrystalline or Rhombohedral
B/B_2O_3 (40 wt.%)		650–950	Au film	Si	30–300	Bundle
N_2/H_2/B_2H_6	N_2	750–1000	Au film	Si (100)	27	Boron nanowires and nanochains
B/B_2O_3 (40 wt.%)		1000–1200	Au film	Si	30–130	Amorphous and Y junctions
B, B_2O_3, C	Ar	1000–1100	Fe_3O_4	Si (001)	20–40	α–tetragonal
B, B_2O_3, Mg	H_2/Ar (5%)	1000–1200	Fe_3O_4	Si (111)	50–200	β-rhombohedral

Table 5.3 Synthesis conditions of boron nanowires using laser ablation methods [5]

Precursor	Carrier gas	T (°C)	Catalyst	Substrate	Diameter	Structure
B target	H$_2$/Ar (5%)	800/ 1300		Si wafer	30–60	Amorphous
B/CoNi target	Ar	1250	CoNi		100	Tetragonal
B, NiCo, H$_3$BO$_3$ target	Ar	1125– 1500	NiCo		<100	Tetragonal
Hot-pressed B pellet	Ar	850– 950	Pt	Sapphire (Cplane)	~100	Crystalline

5.2.1 Chemical Vapor Deposition

CVD is often used to produce high-purity thin films of materials. As the name suggests, the method involves the deposition of volatile precursors that react and/or decompose at the surface of the substrate to produce desired films. The catalyst-assisted CVD technique has been used primarily to synthesize amorphous boron nanowires [3]. The growth of boron nanowires using a similar chemical vapor transportation method have been reported recently [20]. In this methods a mixture of 20–35 mg of boron, 0.5–1 mg of iodine, and 0.5 mg of silicon was put at one end, while a MgO substrate coated with a 5 nm gold thin film was placed at the other end of an evacuated (100 mtorr) sealed quartz tube. The reaction was performed at 1000°C–1100°C, resulting in the production of amorphous boron nanowires with very high aspect ratios, as observed in Fig. 5.3 [20].

Guo et al. have also reported the growth of amorphous boron-rich nanowires on a Ni-coated oxidized Si (111) substrate involving diborane (B$_2$H$_6$, 5 vol.% diluted in H$_2$) as the gas precursor for boron and nitrogen (N$_2$) as the carrier gas at 20 torr and 900°C [35]. The aspect ratio of these nanowires was found to be comparable with those reported by Wu et al. [20].

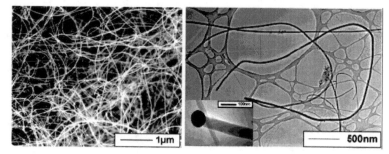

Figure 5.3   (a) SEM image of boron nanowires and (b) TEM image of boron nanowires. The inset in (b) shows the tip of one of the nanowires [20].

Subsequently, this technique has been utilized to grow boron nanowires using the catalyst-free as well as the catalyst-assisted vapor–liquid–solid (VLS) growth method to produce both crystalline and amorphous structures. Otten et al. have reported the growth of crystalline boron nanowires of several micron lengths and diameters ranging from 20 to 200 nm in the form of long straight segments and curly tufts using CVD [15]. These crystalline boron nanowires were synthesized on alumina substrate, where 5% diborane (B_2H_6) in an argon (Ar) gas mixture was passed over a NiB precursor in a fused-silica reactor tube held at 1100°C in a tubular furnace. These nanowires were reported to have a prominent line along the wire axis due to twinning plane (Fig. 5.4a) [15].

Figure 5.4   (a) TEM image showing twinning in a boron nanowire and (b) electron diffraction pattern of the nanowire [15].

The twinning in these crystalline nanowires was verified by the appearance of doubling of spots and parallel streaks between spots and the patterns, as observed in Fig. 5.4b.

Template-assisted synthesis of well-aligned single-crystalline boron nanowires arrays without catalyst using CVD has been reported by Yang et al. [18]. In this method nanochannel alumina (NCA) was used as a substrate and was fabricated using a two-step anodization process in oxalic acid at 40 V for 3 hours and 10 hours, respectively. The NCA so produced was heated at 800°C in a quartz tube furnace. The chamber was pumped down to 20 Pa and a mixture gas of argon, hydrogen, and diborane with a flow ratio of 10:10:1 was filled in the tube furnace. The borane gas is understood to dissociate into boron atom clusters and hydrogen gas (Fig. 5.5a). The boron atoms react with γ-Al_2O_3 of the NCA substrate, forming a Al_5BO_9 layer on the surface of the channels of the NCA, preventing the boron to react with Al_2O_3. The agglomeration of boron clusters confined by the nanochannels is understood to promote the growth of aligned single-crystal boron nanowires in this method.

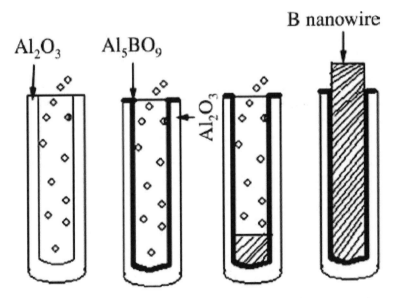

Figure 5.5 Schematic of the growth process of aligned boron nanowires [18].

Yun et al. have fabricated highly inclined bundles of boron nanowire arrays on 5-20 nm thick gold coated silicon substrate by using the oxide-assisted modified VLS approach. In this thermal vapor transport process a mixture of high pure boron with approximately 40 wt.% of B_2O_3 was used as the vapor source in the temperature range of 800°C to 1100°C. The diameter of the boron nanowires were found in the range of 30–300 nm with a large inclined angle as high as 500–600 with respect to the substrate normal [22].

The metal catalyst in this process is employed to generate a solid/liquid (alloy) interface, which plays a critical role in nanowire growth. This enables in production of probable nucleation sites, which initiate the growth of nanowires. Once a solid solution is reached in the solid/liquid alloy, the nanowire growth is initiated. The tip morphology of these nanowires was reported to be temperature dependent, which suggested different growth mechanisms involved in their growth at low and high temperatures, as suggested in Fig. 5.6 [21–23]. These inclined nanowires have been fused to form Y-junctions, which are expected to find potential applications in nanowire functional devices.

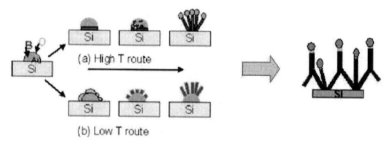

(a) High T route

(b) Low T route

Figure 5.6 Schematic of the growth process of the boron nanowire initiation process (a) at temperatures close to Au-B eutectic temperatures and (b) temperatures much lower than the Au-B eutectic temperatures. The inclined nanowires fuse to form Y-junctions [22].

Methods involving a combination of self-assembly and CVD have been reported to produces highly aligned arrays of boron nanowires. Tian et al. [36] and Liu et al. [24] incorporated the use of self-assembled Fe_3O_4 nanoparticles, which were prepared using high temperature solution phase reaction. The self-assembled assembled layer of these nanoparticles on a silicon substrate acted as a catalyst layer. This provided a good substrate to realize the growth of aligned nanowires using the CVD process.

5.2.2 Magnetic Sputtering

The radio-frequency (RF) MS technique is one of the earliest techniques that have been used to synthesize boron nanowires. Using this technique Cao et al. have grown large-scale self-organized arrays of vertically aligned amorphous boron nanowires having excellent uniformity and high density with diameters 20–80 nm and several tens of micron lengths as observed in scanning electron microscopy (SEM) micrographs (Fig. 5.7) at different magnifications [1, 27].

In this growth process, a mixture of highly pure boron powder and 40 wt.% B_2O_3 powder was used as the target and silicon wafer as the substrate in argon (Ar) gas atmosphere [20]. With the RF power set at 80 W, the temperature of the substrate was maintained at 800°C at 2 Pa pressure during the sputtering process with an argon flow rate of 30 sccm. This led to the deposition of pitch-black film of boron nanowires. It has also been reported that the growth of boron nanowires occurs only at temperatures higher than 700°C under these conditions. Experiments revealed that these nanowires were amorphous in nature and that the choice of substrate had no observable effects on the growth of nanowires using this technique [1, 27, 28, 37].

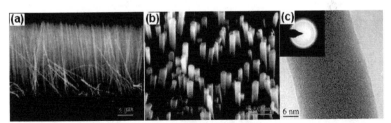

Figure 5.7   (a, b) SEM micrographs of aligned boron nanowires that were grown using the radio-frequency magnetron sputtering technique on a silicon substrate [27]. (c) High-resolution TEM image of a typical boron nanowire. Inset: SAED pattern shows halo rings indicating the amorphous nature of the nanowires [1].

Figure 5.8 shows a schematic model that can be used to understand the growth of amorphous boron nanowires using the RF MS technique. The formation of boron nanowires involves vapor phase generation of the substoichiometric boron oxide (BOx) vapors by sputtering that occurs during the initial stages. The vapor then condenses into BOx nanoclusters by collision of atoms and molecules while drifting toward the substrate. These nanoclusters pile up on the substrate and undergo phase separation of boron and BOx. The oxygen diffuses out, forming an amorphous boron nucleus inside the amorphous BOx sheath. Finally the boron nucleus grows normally to the substrate surface to form boron nanowires. The temperature of the substrate (~800°C) is not high enough to allow crystallization of the boron nanowires during the growth process.

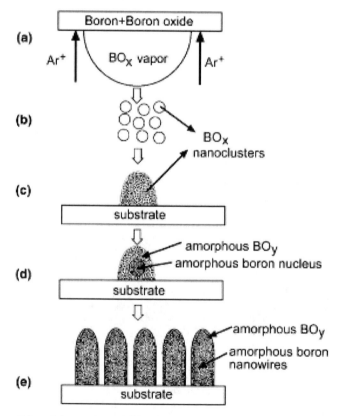

Figure 5.8   Schematic model for the growth of amorphous boron nanowire arrays [38].

The postannealing method has been successfully utilized to prepare crystalline boron nanowires from the amorphous nanowires grown using the RF MS method. The postannealing method [15] involved heating of amorphous nanowires at 1050°C under argon atmosphere in sealed tubes for about three hours. The crystalline boron nanowires so produced when exposed to air gets sheathed by an amorphous oxide layer.

The modified MS method using argon gas has been used to grow multiple Y- or T-junction boron nanowires with unilateral feather-like morphology. These 1D structures fuse together to form feather-like structures (Fig. 5.9) up to several tens of square centimeters [28].

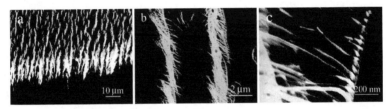

Figure 5.9   SEM micrographs of large-area patterned arrays of boron nanofeathers grown on a silicon substrate [28].

These morphologies were observed as a result of significantly increasing the carrier gas flow rate. Chemical analysis indicates that the nanofeathers produced using a target with different boron powder concentrations were chemically identical. This suggests that the concentration of boron powder in the target does not play any role in the growth of feather-like morphologies. Structural characterizations suggest that the branched boron nanowires preferentially nucleate and grow on the same side wall of the backbone nanowire and align along the same direction to form multiple nanojunctions producing unilateral feather-like morphologies. The nucleation and growth of boron nanofeathers can be controlled by controlling the sputtering conditions. Merely increasing the RF power increases the nucleation density and enables in growth of bushier and highly aligned nanofeathers observed in Fig. 5.10.

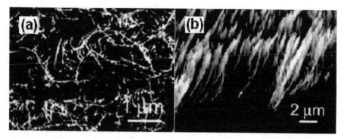

Figure 5.10   SEM micrographs of patterned arrays of boron nanofeathers grown using (a) 40 W of RF power and (b) 120 W of RF power [37].

This feather-like boron nanostructures are amorphous in nature and show diffuse halo rings in SAED patterns. The stem and branches are understood to be coated with an amorphous sheath layer of about 1–2 nm thickness. No other phase precipitation is observed at the junction region between the stem and the branch [29, 37].

5.2.3   Laser Ablation

The LA technique has also been widely used to grow different morphologies of 1D boron nanostructures. Using this method Meng et al. have reported completely catalyst-free growth of boron nanowires [30]. In this process a target was placed at the center of an alumina tube of a high-temperature furnace under a vacuum of 3×10^{-2} torr. This target was made by made by compacting boron powder in a hydraulic press and several silicon strips were placed to act as deposition substrates for grown materials. The system was gradually heated at the rate of 20°C to about 800°C under argon atmosphere mixed with 5% hydrogen flowing at the rate of 50 sccm and a pressure of about 300 torr. After keeping the target at 800°C under these conditions for about 30 minutes, the temperature of the furnace was raised to 1300°C and maintained at this temperature for about 5 hours. The target placed inside the furnace was ablated using a KrF excimer laser at 248 nm with a pulse frequency of 10 Hz [30]. This resulted in the growth of amorphous boron nanowires. The growth of theses amorphous boron nanowires is understood to result from a process combining a vapor-forming agent and a material-transporting process. It thus appears that the growth of these nanowires in the high-temperature LA growth technique is governed by some kind of a vapor–solid (VS) process rather than a VLS process. It has been further observed by Meng et al. that high laser power and high temperatures lead to higher yield of nanowires. Other parameters such as ablation time and pulse frequency have considerable effect, while parameters such as pressure in the tube, flow rate of the carrier gas have no significant effects on nanowire growth. Merely replacing the silicon substrate by a platinum-coated sapphire substrate in this catalyst-free growth methodology has resulted in the growth of crystalline boron nanowires [26].

Transmission electron microscopy (TEM) image analysis shows boron nanowires with a platinum tip, indicating the involvement of the VLS growth mechanism in the boron nanowire growth. It is understood that heating a substrate coated with a very thin film of platinum results in the formation of small nanoislands due to surface tension, as seen in Fig. 5.11. Boron particles formed during the LA process react at the surface of these platinum catalyst islands to form platinum boride, which lowers the melting point of platinum significantly, thus enabling the production of a liquid phase on Pt droplets. The continuous supply of boron migrates into platinum droplets reacting with platinum, resulting in a coexisting system of Pt and PtB. When the dissolved boron reaches the saturation limit it starts to precipitate and accumulate on platinum particles, leading to the growth of boron nanowires. The nanowires produced during this process take β-rhombohedral and α-tetragonal forms, depending on the deposition temperature [26].

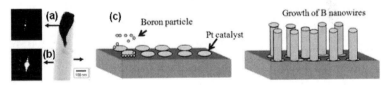

Figure 5.11   (a) TEM image and SAED pattern indicating a platinum tip. (b) TEM image and SAED pattern indicating a boron nanowire. (c) Schematic of the growth mechanism of boron nanowires on a platinum-coated sapphire substrate [26].

Growth of carbon nanotubes by using elements such as Co and Ni as a catalyst has made a significant impact on the growth of boron nanowires using LA methods. Systematic studies on the growth of crystalline boron nanowires by using the LA method, as summarized in Table 5.4, have been reported recently [32, 33]. In these studies the pure boron target was replaced by a target consisting of a boron pellet doped with a combination of different catalyst particles. The impact of laser pulses on the target enabled the growth of nanowires on the target itself. It is, however, observed that the intensity of the laser beam directly affects the morphologies of the product.

The laser energy gets absorbed into the target when impinged upon by a laser pulse. This creates a thermal gradient in the target. The temperature at the core is highest and decreases radially away depending on the thermal conductivity. The effect of this temperature gradient is observed in the density and diameter of the nanowires grown. The density of growth of 1D nanostructure was reported to be more toward the target periphery; however, the diameter became smaller as compared to those near the pores. No growth was reported on the surface of the pore created due to the impact of the laser beam on the target (Fig. 5.12). This happens because the temperature at the core of the pore is very high and the process of ablation occurs continuously, thereby restricting the favorable conditions for the growth.

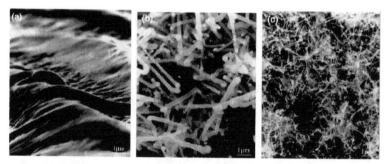

Figure 5.12   SEM images at different positions on the target indicate (a) no growth of nanowires occurs at the surface of the pore, (b) the growth is less at the periphery, while (c) the density increases further away from the pore [32].

The analysis of TEM micrographs shows the presence of catalyst droplets at one end of the nanowires, which suggests the VLS mechanism as one of the growth process involved. Since the

Table 5.4 Summary of target composition and corresponding morphologies obtained using the laser ablation method [32]

No.	B(wt%)	Ni(wt%)	Co(wt%)	W(w/mm^2)	T(°C)	Morphology
1	90	5	5	2	1250	The diameters of the nanorodes or nanowires base on distance from the pore formed by the laser beam
2	90	5	5	5	1250	Mainly powders
3	90	5	5	0.7	1250	Mainly NWs
4	90	5	5	0	1500	Powders
5	90	5	5	2	1200	Low yeild of NWs
6	90	5	5	2	1125	Few NWs
7	90	10	0	2	1250	Similar to Exp. 1
8	90	10	0	2	1200	Similar to Exp. 5
9	90	10	0	2	1125	Similar to Exp. 6
10	90	0	10	2	1250	Fewer NWs than Exp. 1
11	90	0	10	2	1200	Fewer NWs than Exp. 5
12	90	0	10	2	1125	Very few NWs
13[a]	70	5	5	2	1250	Straight NWs
14[b]	100	0	0	2	1250	Boron nanowires with homogenous diameters

[a]The left is H_3BO_3.
[b]The Co catalyst particle with controlled diameters are put near the targert.

targets also contain H_3BO_3, the oxygen-assisted model and/or vapor–cluster–solid (VCS) growth mechanisms are also expected to contribute in the growth of these nanowires [1, 33, 38].

5.3 Properties of Boron Nanowires

Theoretical studies on boron nanostructures predicted that these are quite stable due to the presence of delocalized multicenter bonds, which also play a prominent role in determining their structural and electronic properties [5, 13]. Boron nanowires have been found to exhibit morphology-dependent electrical properties and are known to be semiconducting [4]. Other boron nanostructures such as boron nanosheets, boron nanotubes, and boron fullerenes demonstrate metallic properties [11–13].

5.3.1 Electrical Transport Properties

Otten et al. measured the *I–V* characteristic of three crystalline boron nanowires individually obtained using CVD technique by suspending it between two platinum–iridium probes in a scanning electron microscope [15] and found the nonlinear behavior as shown in Fig. 5.13. They predicted that the nonlinearity in the central region of each curve results from nonohmic electrode contacts with the boron nanowires and is the characteristic of semiconducting nanowires [15, 39, 40].

The electrical conductivity extracted from this *I–V* cure was found to be of the order of 10^{-5} $(\Omega \cdot cm)^{-1}$ compared to conductivity 10^{-4} to 10^{-7} $(\Omega \cdot cm)^{-1}$ of the bulk boron. Since the contact is an important concern while measuring electrical properties, the electrode metal plays an important role in determining the properties. The work function of the electrode material relative to that of the nanowire determines whether the contact is a Schottky or an ohmic contact. Wang et al. measured the conductivity of crystalline boron nanowire devices using nickel and titanium as contact electrodes deposited by electron beam lithography [8]. The work function of intrinsic β-rhombohedral boron is about 4.3 eV and a band gap of about 1.56 eV. The work function of titanium and nickel are known to be

4.33 eV and 5.15 eV, respectively. It was observed that titanium electrodes form a Schottky contact with the boron nanowire, suggesting that the work function of boron nanowires is greater than 4.33 eV. The *I–V* characteristics were found to improve on thermal treatment of the device, suggesting improved contact. The *I–V* characteristics obtained using nickel electrodes suggested an ohmic contact with the boron nanowires. Higher current obtained at negative gate voltages suggests that a boron nanowire is a *p*-type semiconductor, with behavior similar to bulk boron [5, 8]. The *p*-type character of boron can be understood due to the presence of intrinsic structural defects and Jahn–Teller distortions in the unusual icosahedral cluster-based crystal structures.

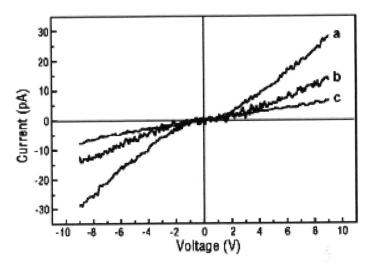

Figure 5.13 *I–V* data for three boron nanowire samples of different diameters show nonlinear behavior in the central region due to nonohmic contact [15].

The effect of mechanical strain on conductivity has been investigated by Tian et al. [4]. Flexible crystalline nanowires were grown using thermoreduction of boron–oxygen compounds with magnesium. Four probe measurements performed using nickel-gold electrodes suggested the electrical conductivity of boron nanowires is ~3.6 × 10^{-2} $(\Omega \cdot cm)^{-1}$. The nanowires were strained, as shown in Fig. 5.14.

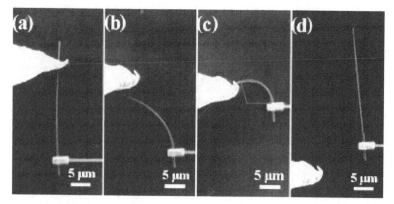

Figure 5.14 SEM micrograph showing the bending process of a boron nanowire [4].

It is observed that the conductance of boron nanowires was very robust and remains almost unchanged even after 3% strain bending. Thus these properties demonstrate that single-crystalline boron nanowires are very promising as building blocks for flexible nanoelectronic devices.

5.3.2 Field Emission Properties

For thermionic electron cathodes in optical instruments such as the transmission electron microscope and the scanning electron microscope, 1D nanostructures such as nanotubes and nanowires have been well accepted as ideal field emission electron sources. A good field emission material requires small electron affinity, high thermal stability, high aspect ratio, and better mechanical strength at high temperatures. Fowler and Nordheim have suggested that nanowires with a low work function will have better field emission performance. These nanostructures help in achieving higher brightness and lower energy spread at a low electric field [41, 42]. One-dimensional nanostructures such as nanotubes or nanowires having sharp tips with diameters ranging from 10 to 100 nm have been found to greatly enhance the local electric field [42–48]. This allows stable field emission of high current density under the above-mentioned practical experimental conditions. The field emitter materials such as carbon nanotubes, diamond films, and transition metal oxide nanowires are limited in building field emission devices due to their poor field emission stability and low melting point [41, 45, 48–51]. One-dimensional boron nanostructures offer many advantages such as high melting point, ensuring long-term stability, low-electron-affinity-induced large field emission current density, and strong mechanical strength compared to other materials [5].

The field emission properties of individual boron nanowires have been studied using a modified high-vacuum scanning electron microscope [24]. In this study a tungsten probe was moved to contact a single boron nanowire and the resistance was recorded to ensure alignment as well as to establish an ohmic contact. The probe was retracted and finally field emission current data was recorded. It is observed that a stable field emission current of 1 μA was reached under the application of an electric field of approximately 59–74 Vμm^{-1} [24]. The maximum current density for boron nanowires was estimated to be in the range of 2×10^5–4×10^5 A/cm^2, indicating that boron nanowires may be used as an excellent high-brightness point electron source. Field emission current data for three different samples in Fig. 5.15a show slightly different field emission properties for different samples. Further investigations by measuring the *I–V* characteristics shown in Fig. 5.15b revealed that nanowires with lower resistance have better field emission properties.

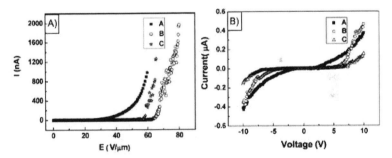

Figure 5.15 (a) The current–electric field (*I–E*) curves and (b) *I–V* characterstics of individual boron nanowires: A, B, and C refer to the three different nanowire samples [4].

The field emission properties of aligned large-scale boron nanowire arrays shows higher current density at lower electric field values without showing any saturation tendency. Since the field emission behavior of boron nanowire arrays was studied on a high-density nanowire film, it partially concealed the true field emission properties of boron nanowires. A clear understanding on the field emission mechanism in boron nanowires is still unclear [5, 24] and is a subject of intense investigation. The field emission behavior of boron nanowire arrays at high current density is different from that of individual boron nanowires. Since the high density of boron nanowire arrays may lead to screening effects during field emission Tian et al. measured the field emission of patterned boron nanowires [36]. The distance between the boron nanowire patterns were about 20–30 μm and each pattern was of 10–15 μm in diameter. They found that the current originates from electron tunneling through the barriers as a result of the electric field and that patterned boron nanowires appear suitable candidates for applications in electron emitting devices.

5.4 Outlook

This chapter discusses in details different synthesis methods that have been used to synthesize crystalline as well as amorphous boron nanowires. Further insight into the processes involved in growth mechanisms using different models, electrical transport, and field emission properties of boron nanowires and parameters affecting them have been developed. Even though several growth models have been developed and used for explaining growth mechanisms using different techniques, a convincing and comprehensive model explaining the growth model and their properties is still unavailable. A major challenge in this area is to develop appropriate theoretical models that will enable in explanation and prediction of growth and properties of these nanowires. This will enable in exploiting these nanowires for application, especially for the development of highly efficient field emission devices.

References

1. Cao, L. M., et al., Template-catalyst-free growth of highly ordered boron nanowire arrays. *Applied Physics Letters*, 2002; **80**(22), 4226–4228.

2. Parakhonskiy, G., et al., High pressure synthesis of single crystals of α-boron. *Journal of Crystal Growth*, 2011; **321**(1), 162–166.

3. Liu, F., et al., Fabrication and field emission properties of boron nanowire bundles. *Ultramicroscopy*, 2009; **109**(5), 447–450.

4. Tian, J., et al., Boron nanowires for flexible electronics. *Applied Physics Letters*, 2008; **93**(12), 122105.

5. Tian, J., et al., One-dimensional boron nanostructures: Prediction, synthesis, characterizations, and applications. *Nanoscale*, 2010; **2**(8), 1375–1389.

6. Cao, L., et al., Large-scale boron nanowire nanojunctions and their highly-oriented arrays. *Science in China Series G: Physics, Mechanics and Astronomy*, 2004; **47**, 621–633.

7. Velamakanni, A., et al., Catalyst-free synthesis and characterization of metastable boron carbide nanowires. *Advanced Functional Materials*, 2009; **19**(24), 3926–3933.

8. Wang, D., et al., Electrical transport in boron nanowires. *Applied Physics Letters*, 2003; **83**(25), 5280–5282.

9. Wang, X. J., et al., Single Crystalline Boron Nanocones: Electric Transport and Field Emission Properties. *Advanced Materials*, 2007; **19**(24), 4480–4485.

10. Tian, J., et al., Boron Nanowires for Flexible Electronics and Field Emission. *AIP Conference Proceedings*, 2009; **1173**(1), 317–323.

11. Lau, K. C., et al., Theoretical study of electron transport in boron nanotubes. *Applied Physics Letters*, 2006; **88**(21), 212111.

12. Lau, K., Y. Yap, and R. Pandey, Boron and boron carbide materials: nanostructures and crystalline solids, in *B-C-N Nanotubes and Related Nanostructures*. Springer New York, 2009; 271–291.

13. He, H., et al., Metal-like Electrical Conductance in Boron Fullerenes. *The Journal of Physical Chemistry C*, 2010; **114**(9), 4149–4152.

14. Yang, Q., J. S., L. Wang, Z. Su, X. Ma, J. Wang, and D. Yang, Morphology and diameter controllable synthesis of boron nanowires. *Journal of Materials Science*, 2006; **41**, 3547–3552.

15. Otten, C. J., et al., Crystalline boron nanowires. *Journal of the American Chemical Society*, 2002; **124**(17), 4564–4565.

16. Chang, M., C. H. L., and J. R. Deka, Characterization and Manipulation of Boron Nanowire inside SEM. *Key Engineering Materials*, 2008; **31–34**, 381–382.

17. Wang, Z., et al., Catalyst-free fabrication of single crystalline boron nanobelts by laser ablation. *Chemical Physics Letters*, 2003; **368**(5–6), 663–667.

18. Yang, Q., et al., Aligned single crystal boron nanowires. *Chemical Physics Letters*, 2003; **379**(1–2), 87–90.

19. Liu, F., et al., Metal-like single crystalline boron nanotubes: synthesis and in situ study on electric transport and field emission properties. *Journal of Materials Chemistry*, 2010; **20**(11), 2197–2205.

20. Wu, Y., B. Messer, and P. Yang, Superconducting MgB2 nanowires. *Advanced Materials*, 2001; **13**(19), 1487–1489.

21. Yun, S. H., et al., Effect of quench on crystallinity and alignment of boron nanowires. *Applied Physics Letters*, 2004; **84**(15), 2892–2894.

22. Yun, S. H., et al., Growth of inclined boron nanowire bundle arrays in an oxide-assisted vapor-liquid-solid process. *Applied Physics Letters*, 2005; **87**(11), 113109.

23. Yun, S. H., et al., Self-assembled boron nanowire Y-junctions. *Nano Letters*, 2006; **6**(3), 385–389.

24. Liu, F., et al., Fabrication of vertically aligned single-crystalline boron nanowire arrays and investigation of their field-emission behavior. *Advanced Materials*, 2008; **20**(13), 2609–2615.

25. Yunpeng, G., Z. Xu, and R. Liu, Crystalline boron nanowires grown by magnetron sputtering. *Materials Science and Engineering: A*, 2006; **434**(1–2), 53–57.

26. Yoon, J.-W., and K. B. Shim, Growth of crystalline boron nanowires by pulsed laser ablation. *Journal of Ceramic Processing Research*, 2011; **12**(2), 199–201.

27. Cao, L. M., et al., Well-aligned boron nanowire arrays. *Advanced Materials*, 2001; **13**(22), 1701–1704.

28. Cao, L. M., et al., Featherlike boron nanowires arranged in large-scale arrays with multiple nanojunctions. *Advanced Materials*, 2002; **14**(18), 1294–1297.

29. Wang, Y. Q., L. M. Cao, and X. F. Duan, Amorphous feather-like boron nanowires. *Chemical Physics Letters*, 2003; **367**(3–4), 495–499.

30. Meng, X. M., et al., Boron nanowires synthesized by laser ablation at high temperature. *Chemical Physics Letters*, 2003; **370**(5–6), 825–828.

31. Zhang, Y., et al., Synthesis of crystalline boron nanowires by laser ablation. *Chemical Communications*, 2002; (23), 2806–2807.

32. Zhang, Y., et al., Study of the growth of boron nanowires synthesized by laser ablation. *Chemical Physics Letters*, 2004; **385**(3–4), 177–183.

33. Guo, T., et al., Catalytic growth of single-walled nanotubes by laser vaporization. *Chemical Physics Letters*, 1995; **243**, 49–54.

34. Wang, N., et al., Nucleation and growth of Si nanowires from silicon oxide. *Physical Review B*, 1998; **58**(24), R16024–R16026.

35. Guo, L., R. N. Singh, and H. J. Kleebe, Growth of boron-rich nanowires by chemical vapor deposition (CVD). *Journal of Nanomaterials*, 2006; **2006**, 6.

36. Tian, J., et al., Patterned boron nanowires and field emission properties. *Applied Physics Letters*, 2009; **94**(8), 083101.

37. Cao, L. M., et al., Nucleation and growth of feather-like boron nanowire nanojunctions. *Nanotechnology*, 2004; **15**(1), 139.

38. Wang, Y. Q., et al., One-dimensional growth mechanism of amorphous boron nanowires. *Chemical Physics Letters*, 2002; **359**(3–4), 273–277.

39. Chung, S.-W., J.-Y. Yu, and J. R. Heath, Silicon nanowire devices. *Applied Physics Letters*, 2000; **76**(15), 2068–2070.

40. Cui, Y., et al., Doping and electrical transport in silicon nanowires. *The Journal of Physical Chemistry B*, 2000; **104**(22), 5213–5216.

41. Li, Y. B., et al., Field emission from MoO_3 nanobelts. *Applied Physics Letters*, 2002; **81**(26), 5048–5050.

42. Zhang, H., et al., Field Emission of electrons from single LaB6 nanowires. *Advanced Materials*, 2006; **18**(1), 87–91.

43. Wong, K. W., et al., Field-emission characteristics of SiC nanowires prepared by chemical-vapor deposition. *Applied Physics Letters*, 1999; **75**(19), 2918–2920.

44. Chen, J., et al., Field emission from crystalline copper sulphide nanowire arrays. *Applied Physics Letters*, 2002; **80**(19), 3620–3622.

45. Li, Y. B., Y. Bando, and D. Golberg, ZnO nanoneedles with tip surface perturbations: Excellent field emitters. *Applied Physics Letters*, 2004; **84**(18), 3603–3605.

46. Lee, Y.-H., et al., Tungsten nanowires and their field electron emission properties. *Applied Physics Letters*, 2002; **81**(4), 745–747.

47. Yin, L. W., et al., Growth and field emission of hierarchical single-crystalline wurtzite AlN nanoarchitectures. *Advanced Materials*, 2005; **17**(1), 110–114.

48. Chernozatonskii, L. A., et al., Electron field emission from nanofilament carbon films. *Chemical Physics Letters*, 1995; **233**(1–2), 63–68.

49. Wang, W. Z., et al., Aligned ultralong ZnO nanobelts and their enhanced field emission. *Advanced Materials*, 2006; **18**(24), 3275–3278.

50. Geis, M. W., et al., Electron field emission from diamond and other carbon materials after H2, O2, and Cs treatment. *Applied Physics Letters*, 1995; **67**(9), 1328–1330.

51. de Heer, W. A., A. C., and D. Ugarte, A carbon nanotube field-emission electron source. *Science in China Series G: Physics, Mechanics and Astronomy*, 1995; **270**(5239), 1179–1180.

Chapter 6

Applications of Boron Nanostructures in Medicine

Komal Sethia and Indrajit Roy
Department of Chemistry, University of Delhi, Delhi 110007, India
Indrajitroy1@gmail.com

6.1 Introduction

Over the past few years, several inorganic-based nanoparticles have emerged as promising candidates for various biomedical applications. Their structural robustness, resistance to microbial attack, nonimmunogenicity, flexible chemistry, etc., are key aspects that highlight the importance of inorganic nanoparticles in medicine [1–3]. Interestingly, the applications of such nanoparticles are quite distinct from those of their polymeric or lipidic counterparts. The later types of particles, such as polymeric micelles, polyester nanoparticles, solid–lipid nanoparticles (SLPs), etc., have been primarily used in the controlled release of encapsulated drug molecules [4–6]. On the other hand, inorganic-based nanoparticles have largely emerged as diagnostic probes (e.g., iron oxide nanoparticles, quantum dots [QDs], gold nanoparticles, etc.),

Handbook of Boron Nanostructures
Edited by Sumit Saxena
Copyright © 2016 Pan Stanford Publishing Pte. Ltd.
ISBN 978-981-4613-94-1 (Hardcover), 978-981-4613-95-8 (eBook)
www.panstanford.com

antimicrobials (e.g., silver nanoparticles), and, more recently, probes for externally activated therapies (e.g., gold nanoshells) [1, 7–9].

This book chapter focuses on the applications of boron-based nanostructures in medicine. Such nanostructures involve pure boron nanoparticles (BNPs), boron nitrides/carbides, boron nanotubes (BNTs), boron-containing nanoheterostructures such as boronated porphyrins, magnetically doped BNPs, etc. Though such nanostructures have a variety of applications in medicine, they are specifically suited for unique externally activated therapy, namely, boron neutron capture therapy (BCNT). This chapter will begin with a discussion about BCNT. This will be followed by detailed discussions about the various boron nanostructures, covering their synthesis and characterization, biocompatibility and aqueous dispersion, surface modification and various biomedical applications (diagnosis, therapy, biosensing, etc.). The chapter will conclude with discussions about the future prospects of such nanomaterials in medicine.

6.2 Boron Neutron Capture Therapy

These days, externally activated therapies are being envisioned as the future of medicine, mainly cancer therapy [9, 10]. In such therapies, a nontoxic probe (sensitizer) is transformed into a toxic form following irradiation with an external source, such as visible/near-infrared (NIR) light, a magnetic field, X-rays, or thermal neutrons. These sensitizers are first targeted to diseased sites, even though off-target accumulation cannot be completely avoided. However, since the subsequent external radiation is focused only at the diseased site, the toxicity is limited only to the target site, with the probe remaining nontoxic at other, healthy sites. Therefore, such therapies promise to completely avoid off-target toxicity and therapy-related side effects, thus overcoming the major drawbacks associated with conventional therapies.

Neutron capture therapy (NCT) was first suggested in 1936 by Locher [11]. BNCT relies on the selective accumulation of boronated sensitizers within tumor tissue, followed by their activation upon irradiation with low-energy neutrons [12]. Specifically, BNCT is based on the nuclear capture and fission reactions that occur when boron-10 (10B), nonradioactive isotope of boron, is irradiated with low-energy (0.025 eV) thermal neutrons. This resulted in the production of high-linear-energy-transfer (LET) alpha particles (4He) and recoiling lithium-7 (7Li) nuclei, as shown in Fig. 6.1 [13].

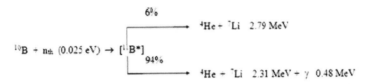

Figure 6.1   Scheme showing the production of high-linear-energy-transfer (LET) alpha particles (4He) and recoiling lithium-7 (7Li) nuclei following neutron bombardment of a boron-10 nucleus [13].

BNCT using a single drug has various advantages, including increased therapeutic effect due to the targeting of different cellular components and/or mechanisms of tumor cell destruction. Despite their high promise, there are various major challenges involved in the development of boron delivery agents for BNCT, which have to be overcome before they can be commercialized:

- Achieving boron concentrations of 109 of 10B per cell
- Normal tissue uptake
- With a tumor: normal tissue and tumor: blood (T:Bl boron concentration ratios of ~3); tumor boron concentration of (~20 µg 10 B/g tumor)
- Low toxicity
- Water solubility
- Chemical stability
- Relatively rapid clearance from blood and normal tissues
- Persistence in tumor during neutron irradiations [12, 14]

The only two BNCT delivery agents currently used in clinical trials are nanoparticles of sodium mercaptoundecahydro-closo-dodecaborate (Na2B12H11SH), commonly known as sodium borocaptate (BSH), and a macromolecule of boron-containing amino acid (L)-4-dihydroxy-borylphenylalanine, known as boronophenylalanine, or BPA. But neither of them fulfills the criteria indicated above [14], which necessitates the synthesis of new generation of boronated probes. More recently, nanoparticles have emerged as promising boron-containing probes owing to a number of positive attributes. Nanoparticles can achieve higher boron concentration within a single probe and render them with stable aqueous dispersibility. Greater tumor selectivity of the probes can be achieved via the well-known enhanced permeability and retentivity (EPR) effect and/or the inclusion of tumor-targeting molecules, such as peptides, proteins, antibodies, nucleosides, sugars, porphyrins, liposomes, and nanoparticles [1]. Moreover, probes for other imaging and/or therapeutic modalities can be co-incorporated within boronated nanoparticles, which may allow multimodality and *theranostics*. Recently, boron-containing silica [15], carbon [16], and silver [17] nanoparticles, as well as boron carbide nanoparticles [18], have also been reported as potential boron delivery agents with high boron content.

6.3 Various Boron-Containing Nanoparticles in Medicine

6.3.1 Pure Boron Nanoparticles

Pure BNPs are superior boron delivery agents with very high boron density (ca. 107 boron atoms per 50 nm particle). However, hydrophilic BNPs with well-defined surface properties have not yet been reported [14]. The synthesis of BNPs using methods such as gas-phase pyrolysis of diborane or solution-based synthetic routes have been reported [19–21]. However, these methods have inherent problems of complicated and hazardous reaction conditions, low yield, along with poor stability of resulting particles. Recently, a surfactant-assisted ball-milling method used to produce air-stable BNPs on a large scale has been reported [22, 23]. This method is simple, inexpensive, and easily scalable. The particles synthesized by this method are of approximately 50 nm in size and no primary particles were found that were larger than 100 nm in size. The surfactant-assisted ball-milling method was initially designed to produce oxide-free, hydrophobic BNPs that are stabilized by fatty acid ligands, for example, undecylenic acid (UND). Their stable and nonaggregated aqueous dispersion was achieved by a subsequent ligand exchange process, whereby the hydrophobic ligands were replaced by hydrophilic bidentate ligands with amine functionality, such as dopamine (DA) and polyethylene glycol (PEG) amine [14]. In addition to stable aqueous aggregation, the terminal amine groups also provided sites for further chemical modification and conjugation with other bioactive molecules or contrast agents [22].

In a recent report, Gao et al. have prepared DA-BNPs with an average diameter of 40 nm [14]. These nanoparticles were water dispersible and their aqueous suspensions were collidally stable for over a month. Moreover, their surfaces presented free amino groups, which can be further used for conjugation with fluorophores and/ or drugs. The DA-BNPs did not show any sign of cytotoxicity when treated with murine macrophage cells (RAW264.7). These results suggest that DA-BNPs, once conjugated with cancer-targeting ligands, may provide us with unique boron-rich carriers for BCNT.

6.3.2   Boron-Containing Nanoparticles

Literature pertaining to the use of pure BNPs in medicine is scarce. However, plenty of reports exist regarding the use of nanoparticles containing boron as a constituent, whether they can be either boron compounds or nanoclusters of boron with other elements or organic/polymeric heterostructures containing boron. In this section, we discuss the synthesis and biological applications of some inorganic nanoparticles containing boron as a constituent.

6.3.2.1   Boron carbides

Boron carbide is one of the hardest materials known. The research on the synthesis of nanosized boron carbide is of great importance [24]. Solid boron carbide nanoparticles possess high concentration of bulk boron atoms in comparison to molecular formulations. In addition, boron carbide is very inert and not easily subjected to degradation, which is an essential requirement for BNCT. Boron carbide nanoparticles can be produced by radio-frequency (RF) plasma synthesis or the ball-milling method in an argon or nitrogen atmosphere from commercially available boron carbide [18, 25], with the ball-milling method being the most common method.

Amorphous-like boron carbide nanoparticles were prepared using the ball-milling method, and according to photon correlation spectroscopy, a size distribution of 73 nm was obtained [18]. The chemical analysis confirmed that milling in different types of atmosphere can alter the surface chemistry. For example, milling in an air and nitrogen atmosphere results in introduction of new oxygen- and nitrogen-containing species, whereas milling in an inert (argon) atmosphere results in pure boron carbide nanoparticles. This altered surface chemistry was directly observed via change of electrophoretic mobility in agarose gel electrophoresis. An in vitro experiment revealed significant cell death upon thermal neutron irradiation of cells treated with boron carbide nanoparticles, whereas irradiation alone or treatment with nanoparticles without irradiation did not cause any cellular toxicity. These results showed the feasibility of boron carbide nanoparticles as a new agent for BNCT [18].

The surface modification of boron carbide nanoparticles is based on both covalent binding to the particle surface and as- sociation by hydrophobic interaction, which allowed their bio- conjugation with the transacting transcriptional activator (TAT) peptide and fluorescent dyes. Mortensen et al. used the amino reactive fluorophore, rhodamine B sulfonyl chloride (lissa- mine), to fluorescently label the nanoparticles. The bifunction- al linker 4-(N-maleimidomethyl)cyclohexane-1-carboxylic acid N-hydroxysuccinimide (SMCC) was used to couple a cysteine-con- taining peptide sequence (HS-TAT-FITC, the TAT peptide sequence being labeled with a fluorescein isothiocyanate [FITC] fluorophore) to a surface-bound amino group. These coated nanoparticles can be easily translocated into murine EL4 thymoma (EL4 T) cells and B16 F10 malignant melanoma cells in amounts as high as 0.3 wt.% and 1 wt.%, respectively. Figure 6.2 shows the uptake of fluores- cently labeled boron carbide nanoparticles in EL4 T cells via con- focal microscopy. In vitro BNCT experiments showed that neutron irradiation caused not only the death of cells loaded with boron- containing nanoparticles but also the death of neighboring, non- loaded cells. These results show that functionalized boron carbide nanoparticles may be used for BNCT as both initial high loading of cells with 10B and the effective killing of loaded and neighboring unloaded cells are achieved [25]. Subsequent in vivo experiments in mice bearing subcutaneous tumors, generated from aggressive B16-OVA melanoma cells, reflected the in vitro data. Here, the com- bination of treatment with BNPs and exposure to thermal neutron radiation significantly delayed the tumor growth and prolonged the survival of tumored mice when compared to radiation-exposed tumored mice not treated with the nanoparticles [26]. The results showed the effectiveness of boron carbide nanoparticles as probes for BNCT, both in vitro and in vivo.

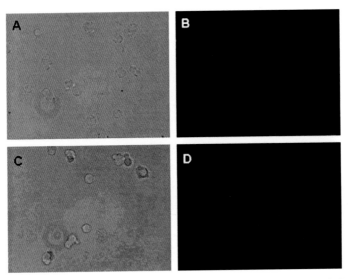

Figure 6.2   Bright-field (left) and fluorescence microscopy (right) of EL4 T cells loaded with fluorescently labeled boron carbide nanoparticles (A and B) and cells alone (C and D) [25].

6.3.2.2 Boron nitride nanotubes

Boron nitride nanotubes (BNNTs) are probably the most versatile of all boron-containing nanostructures, being the structural analogues of carbon nanotubes (CNTs) with alternating B and N atoms [27]. BNNTs are traditionally synthesized using a combination of ball-milling and annealing methods [28, 29]. However, owing to the high temperature requirements of this technique, newer techniques requiring lower synthesis temperature of BNNTs, such as catalytic chemical vapor deposition (CCVD) and plasma-enhanced pulsed-laser deposition (PE-PLD), are fast gaining popularity [30]. A layered multiwalled BNNT structure is chemically more stable than a graphitic carbon structure, which also attributes for their poor dispersibility in the aqueous solvents, thereby making them unsuitable for biological applications. Since BNNTs are composed of a significant number of boron atoms per nanotube, they can be explored as vectors of boron atoms for BCNT [31]. The problem of their poor aqueous dispersibility can be overcome using a technique of noncovalent wrapping of polyelectrolytes, such as polyethyleneimine (PEI) and poly-L-lysine (PLL), on the BNNTs. This allows for the aqueous dispersion of BNNTs, thus facilitating their biological applications [32].

Ciofani et al. carried out a study on the cytocompatibility and in vitro cell uptake efficiency of PEI-coated BNNTs. Ball milling and subsequent annealing with anhydrous NH3 were used to prepare the BNNTs, which were further coated with PEI for stable aqueous dispersions. The results showed that the viability of human neuroblastoma (SH-SY5Y) cells were not affected upon interaction with PEI-BNNTs up to a concentration of 5 $\mu g\ mL^{-1}$ PEI-BNNTs in the cell culture medium [32]. The authors followed up this preliminary work by incorporating fluorescent QDs with the PEI-BNNTs, which allowed intracellular fluorescent tracking of the probes. Cell uptake studies were done by studying the uptake of carboxyl QD–conjugated BNNTs into human neuroblastoma SH-SY5Y cells. It was shown that the viability and metabolic activity of the cells remained unaffected even after 72 hours of treatment with the QD-PEI-BNNTs. Fluorescence microscopy studies demonstrated that the internalization of QD-PEI–BNNTs follows an energy-dependent endocytotic pathway [33].

The identification of magnetic properties of BNNTs added to their versatility and promise of potential applications in medicine via magnetically guided drug delivery, where magnetic nanoparticles are concentrated at target cells/tissues using an externally applied magnetic field [34]. The reason for the magnetic properties in BNNTs is mainly due to the presence of small Fe particles as production residues, which acts as catalysts to assist nanotube growth during the annealing process [35]. Figure 6.3 shows the transmission electron microscopy (TEM) image of a multiwalled BNNT containing a Fe catalyst particle at one tip. Superconducting quantum interference device (SQUID) analysis has confirmed their superparamagnetic character, whereas other important magnetic parameters, such as magnetic permeability and magnetic momentum, were identified by energy-dispersive X-ray spectroscopy (EDS) and TEM studies. Under the influence of an external magnetic field, QD-conjugated BNNTs were significantly internalized by SH-SY5Y cells, as shown by fluorescence microscopy. The uptake was shown to be dependent on the distance between the magnet and cells, the lesser distance showing higher uptake. The synthesized BNNTs not only showed considerable cell internalization but also potential use as magnetically driven nanovectors for drug delivery [35].

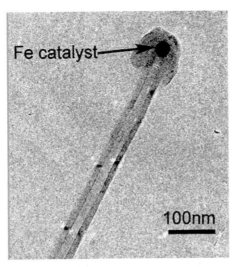

Figure 6.3   TEM image of a multiwalled BNNT containing a Fe catalyst particle at one tip [35].

The piezoelectric properties of BNNTs make them attractive candidates as bionanotransducers for cell sensing and stimulation and may have important applications in gene delivery [36]. Present methods of gene delivery either rely on potentially hazardous engineered viruses or synthetic nonviral nanocarriers to transport therapeutic genes to target tissues and subsequently inside cells [37]. However, slow gene accumulation and poor concentration in target cells, when delivered through nonviral carriers, are major blocks that undermine the efficacy of gene transfer [38]. Though surface functionalization of nonviral carriers with biotargeting agents can marginally enhance gene accumulation in target tissues, they often fail to achieve significant target accumulation and gene concentration [38, 39]. BNNT-mediated enhanced electropermeabilization of cells with low electric fields can significantly enhance the efficacy of gene transfer in target cells, as shown by Raffa et al. [40]. This study aimed at resolving the problems associated with the high voltage required in clinical electropermeabilization. The BNNTs were synthesized via ball milling and annealing with Fe and Cr as catalysts and PLL as a dispersing agent. They found a cell permeabilization efficiency of 40% at 40 V cm^{-1} and 80% at 60–80 V cm^{-1}, with treated cells remaining healthy and metabolically active. The BNNTs acted as mediators, lowering the actual electric field requirement of 300 V cm^{-1}, for repairable cell permeabilization [40].

Along similar lines, Huang et al. demonstrated that BNNTs coated uniformly with Fe_3O_4 nanoparticles enabled these nanocomposites to respond well and be easily physically manipulated in a low magnetic field. The superparamagnetic Fe_3O_4 nanoparticles were uniformly assembled and immobilized via an ethanol-thermal process [41]. The well-documented method surface functionalization for enhanced gene delivery has also been adopted in the case of BNNTs, thus mediating their cell-binding properties. Xing Chen et al. have shown that pristine multiwalled BNNTs of outer diameter 20–30 nm were synthesized via a chemical vapor deposition process. These BNNTs were functionalized with glycodendrimers and were demonstrated to be successful carriers of DNA oligomers for intracellular delivery, displaying no cytotoxicity [42]. Significant cell viability and metabolic activity was observed, which were further confirmed by similar conclusions from DNA content measurement using the PicoGreen dye and the LIVE/DEAD viability/cytotoxicity kit [43]. However, BNNTs' foray into the biomedical area has necessitated detailed studies into their cytocompatibility. Several in vitro studies have been carried out on a number of cell types like Chinese hampster ovary (CHO) cells and human embryonic kidney (HEK 293) cells, with the results showing optimal cytocompatibility [40]. An in vivo pilot toxicological study of BNNTs was done in New Zealand rabbits, using chitosan-coated BNNTs up to concentrations of 1 mg of nanoparticles per kilogram of body weight of animals. The results obtained by Ciofani et al., after 72 hours, are promising and showed no negative effect of BNNTs on blood parameters [44].

Another very unique area of potential application of BNNTs is in electrostimulation of cells/tissues, without the requirement of cumbersome electrode implantation and connection to external electric source. Piezoelectric BNNTs have been shown to act as nanotransducers for electrical/mechanical signals within cellular environments. Externally activated ultrasound was utilized by Ciofani et al. to impart mechanical stress to the BNNTs internalized by neuronal-like PC12 cells. The polarized nanotubes conveyed electrical signals to the cells, stimulating them to display a pronounced outgrowth of neutrites in PC12 cultures, as shown in Fig. 6.4 [45].

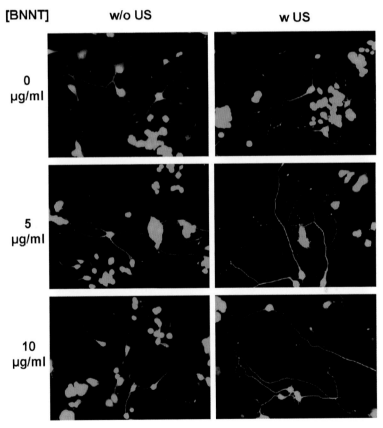

Figure 6.4 Images of calcein-labeled PC12 cells after nine days of treatment (with and without BNNT incubation; with and without ultrasound [US] stimulation). Magnification 20× [45].

6.3.2.3 Nanoclusters of boron with iron

Another type of magnetically doped boron nanomaterial is nanoclusters of iron oxide and boron. Ciobanu demonstrated that it is possible to prepare biocompatible iron oxide–boron nanoclusters via wet chemistry using the nontoxic polyol propylene glycol as the capping agent. Nanoclusters with sizes ranging from 5 nm to 200 nm were synthesized with tunable boron content and showed impressive stability in physiological solutions. The synthesized nanoclusters showed great potential as magentically targeted drug delivery systems in cancer therapy where chemotherapeutics could be loaded into the nanoclusters and magnetically targeted to the tumor site, where the therapeutic payload will be released, with the additional provision of concurrent or subsequent BNCT [46].

6.3.2.4 Boron–organic hybrid nanoparticles

Various organic units (porphyrins, chlorins, bacteriochlorins, polymeric micelles etc.) can be linked to boron atoms/molecules to give complex metal–organic heterostructures with nanosized dimensions. These nanohybrids have number of medical applications, such as BNCT, photodynamic therapy (PDT), gene delivery, diagnostics, etc. [1, 47]. Conjugated boron-containing molecules with porphyrin derivatives can benefit from the recognized ability of many porphyrins to accumulate efficiently in cancer cells [48]. The Sonogashira reaction is mainly used for the synthesis of boron containing nanosized conjugates of natural porphyrins, chlorins, and bacteriochlorins, etc. [49]. Porphyrin-assisted enhanced boron delivery in cancer cells is expected for these kinds of conjugates. Also, several porphyrin derivatives are photosensitizers, which serve as probes for another type of externally activated therapy, namely, PDT. The other intrinsic property of several porphyrins is their fluorescence, which makes them useful optical probes. Therefore, most boronated porphyrins (macromolecules) not only are intrinsically suitable for both BNCT and PDT but also can serve as optical contrast agents.

Boron nanostructures such as carboranes and metal bis(dicarbollide) complexes are highly capable of loading a porphyrin derivative with up to 18 boron atoms per nanoparticle

[50]. The porphyrin fragment in these compounds facilitates the targeting of boron to cancer cells. Porphyrin derivatives called chlorins absorb light in the 660 nm region, fluoresce at 670 nm and have high quantum yield of singlet oxygen generation. Therefore, chlorin chromophores (macromolecules) are highly suitable for creation of boronated conjugates. Chlorins conjugated with BNPs such as o-carboranes, 1-carba-closododecaborate anions, nido-carborate anions, and mercapto-closo-dodecaborate anions were also reported to penetrate in cancer cells and possess photoinduced cytotoxicity, with the added possibility of BNCT [51–55].

Efremenko et al. have synthesized boron cluster nanoparticles via complex conjugates of boron with chlorin e6 using a flexible diethylene glycol linker [56]. The fluorescent properties of these nanoparticles made them excellent optical probes. Fluorescence microscopy revealed that these nanoparticles could infiltrate A549 human lung adenocarcinoma cells, diffusely stain cytoplasm, and accumulate highly in lysosomes, while remaining nontoxic. Average cytoplasmic concentration of boron atoms (B) reached 270 µM (ca. 2×108 B/cell). Such nanoparticles were found to cause photoinduced cell death at micromolar concentrations, with high quantum yields of singlet oxygen generation (up to 0.85 in solution). However, their BNCT efficacy is yet to be studied [56].

6.3.2.5 Boron nanoparticles associated with polymeric micelles

A drug delivery system which is based on nanomaterials, including proteins, drug-conjugated polymers, and nanosized particles, enabled the accumulation of drugs in the tumor region because of the so-called EPR effect [57, 58]. To improve the accumulation efficiency of nanoparticles in the tumor region by the help of EPR effect, it is essential that the drug-carrying nanoparticles circulate for a longer time in the blood without any leakage of drug molecules during circulation. It has been demonstrated that the presence of inert polymers, such as PEG and dextran, on the surface of nanoparticles can enhance their circulation time in the blood [1]. The EPR effect is also expected to play a major role in nanoparticle-assisted BNCT of cancer. Therefore, PEG-grafted nanoparticulate carriers, such as liposomes and dendrimers, have been investigated for the delivery of encapsulated boron compounds in the tumor. However, the premature leakage of the boron-based drugs from their nanocarriers prevented satisfactory boron accumulation in the tumors [59].

To overcome the problem of premature drug release, Sumitani and Nagasaki have synthesized polymeric micelles covalently linked with boron compounds by the free-radical copolymerization of PEG and polylactic acid (PLA), in the presence of polymerizable boron clusters (carboranes containing vinylbenzene groups). The resulting PEG-b-PLA block polymers, bearing an acetal group at the PEG end and a methacrylic (MA) group at the PLA end (acetal-PEG-b-PLA-MA), forms polymeric micelles in an aqueous medium, with the PLA-MA groups forming the inner hydrophobic core. The carborane vinylbenzene monomer dissolves into the hydrophobic core of the micelles, with the vinylbenzene groups and methacrylic groups forming a polymeric network [59]. Thus, the boron-containing carborane moieties get covalently incorporated within the micelles, instead of being merely physically entrapped. The obtained boron-conjugated micelles showed extremely high stability under physiological conditions, with negligible release of boron from the core due to the presence of covalent bonds. While this nonrelease of the drug will be an impediment for conventional therapy, in the case of BNCT, no release of the boronated agents from their carrier micelles is required. The irradiation of the boronated agents by thermal neutrons can take place even within the covalently conjugated micelle, and cellular damage can still be caused by the generated alpha particles and lithium. The PEG corona of these micelles ensured their physiological stability and long circulation in the blood. In vivo experiments were conducted by intravenous injection of these boronated micelles in tumor-bearing mice, which demonstrated robust tumor accumulation (7.2 ID % per g) in tumors, owing to the long circulating nature of the micelles and the EPR effect. Subsequent thermal neutron irradiation resulted in significant suppression of the growth of tumor in tumor-bearing mice treated with the boronated micelles. Moreover, the boronated micelles administered via intravenous injection excreted easily from major organs (except tumor) via biodegradation of the PLA core. Therefore, it can be concluded that the boron-conjugated micelles are expected to exert significant therapeutic effects and can provide a promising platform for the creation of novel boron carriers for cancer BNCT [59].

6.4 Other Applications of Boron Nanoparticles

6.4.1 Boron Nanoparticles in Tissue Engineering

Owing to their structural robustness, boron-based nanostructures have also found applications in the field of tissue engineering and scaffold engineering. Lahiri et al. have studied the potential of BNNTs as reinforcement in orthopedic scaffolds based on polylactide–polycaprolactone copolymers. Orthopedic scaffolds are required to have very good mechanical properties and biocompatibility. Reinforced scaffolds showed improved tensile strength and elasticity, without any cytotoxic effects on osteoblasts and macrophages. The enhanced mechanical properties and biocompatibility of the BNNT-reinforced scaffold has made it a suitable candidate as an orthopedic scaffold. The synthesized scaffolds also regulated osteoblast differentiation [60].

6.4.2 Boron Nanoparticles in Biosensing

Nanoscale biosensors have gained tremendous potential in recent times owing to the possibility of packing a larger number of sensing elements onto a smaller device by virtue of the nanodevice's large surface-to-volume ratio. The requirement for smaller volumes of test materials and the ability to functionalize them with biological molecules of interest also make these sensors very attractive, not to mention their higher sensitivity, selectivity, and longer life span [61]. Chowdhury et al. utilized BNNTs as mass sensors in two different configurations, namely, cantilevered and bridged, as depicted in Fig. 6.5. They measured the amount of biomass by measuring the shift in the resonance frequency. Their study brought them to the conclusion that BNNT-based biosenors could show a sensitivity of 0.1 zg/GHz with the cantilever configuration trumping over the bridged configuration, with a fourfold increase in sensitivity [62].

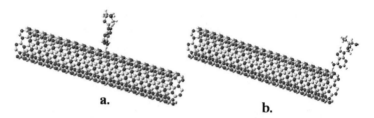

Figure 6.5 Carton of (a) bridged BNNT biosensor and (b) cantilever BNNT biosensor [62].

Polyfluorene nanofilms containing boron have been reported to behave as ultrasensitive biosensors to detect trace amounts of fluorine in aqueous solutions using the nondestructive and highly sensitive technique of impedence spectroscopy [63]. The decoration of thermal oxidation-resistant boron nitride nanosheets with silver nanoparticles for the preparation of robust and reusable nanohybrids for surface-enhanced Raman scattering (SERS)-based sensors was attempted by Lin et al. This SERS-based sensor shows promise in selective detection of biological molecules [64]. Ni-encapsulated BNNTs have been reported as excellent candidates for optomagnetic-based sensor devices owing to the inherent photoluminescent property of the nanotubes [65]. Biocompatible and luminescent difluoroborondibenzoylmethane-polylactide nanoparticles have been shown to exhibit fluorescence, two-photon absorption, and stable phosphorescence at room temperature. The study conducted by Pfister et al. has proved that not only are they potential biosensors but they can also be used as imaging agents, as given by the internalization of these nanoparticles by CHO cells [66]. Multiwalled BNNTs have been functionalized with biotin-fluorescein with anchored Ag nanoparticles and have been studied for use as 3D pH-mapping probes, in addition to fluorescence measurement. The proposed ratiometric sensor provides the capability of in vivo spatially resolved pH mapping in living cells. Using these sensors Huang et al. were able to successfully carry out SERS measurements and a 2D pH distribution map of NIH/3T3 cells [67].

6.5 Conclusion

With the demonstration of their nontoxicity, along with their increased tumor-specific accumulation, boron-containing nanoparticles are expected to play a bigger role in medicine in the future. The last two decades in nanomedicine research have witnessed a plethora of polymeric drug nanocarriers and inorganic diagnostic nanoprobes, coupled with biorecognition agents for disease-specific delivery. However, the outcomes of such research have been largely disappointing, with unsatisfactory diseased/ normal tissue accumulation, and triggering of immunogenicity upon incorporation of protein-/peptide-based targeting agents. As a result, delivery of highly cytotoxic drugs using the above nanocarriers is yet associated with poor therapeutic benefits and severe side effects. Therefore, the concept of externally triggered therapies at diseased sites using nontoxic precursor agents is being visualized to provide enhanced therapeutic gain. The nontoxicity and high tumor-specific boron accumulation achieved by several boron-containing nanoparticles have generated a renewed enthusiasm in BNCT, which was previously thought to be ineffective when boronated molecules were used as probes. In addition, incorporation of diagnostic probes such as iron oxide nanoparticles and QDs in these carriers impart them with diagnostic ability, which would be beneficial in labeling of diseased tissues prior to neutron irradiation. Boronated nanoprobes can also be incorporated with conventional chemotherapeutic agents and genetic therapeutics that may synergize with BNCT. In addition, the mechanical and electrochemical properties of BNNTs are expected to play a critical role in tissue engineering, prosthetics, and biosensors. Taken together, boron-containing nanoparticles are highly promising from the point of view of medical applications and therefore deserve detailed and extensive investigations in the laboratory, preclinical, and clinical setup.

References

1. Prasad P. N., *Introduction to Nanomedicine and Nanobioengineering* (Wiley, New York, 2012).

2. Jain T. K., Roy I., De T. K., and Maitra A. N., *Journal of the American Chemical Society*, **120**, 11092–11095 (1998).

3. Yong K. T., Roy I., Swihart M. T., and Prasad P. N., *Journal of Materials Chemistry*, **19**, 4655–4672 (2009).

4. Farokhzad O. C., and Langer R., *ACS Nano*, **3**(1), 16–20 (2009).

5. Cabral H., Nishiyama N., and Kataoka K., *Accounts of Chemical Research*, **44**(10), 999–1008 (2011).

6. Musacchio T., and Torchilin V. P., *Frontiers in Bioscience*, **16**, 1388–1412 (2011).

7. Huang X., Jain P. K., El-Sayed I. H., and El-Sayed M. A., *Nanomedicine (London)*, **2**(5), 681–693 (2007).

8. Smith A. M., Duan H., Mohs A. M., and Nie S., *Advanced Drug Delivery Reviews*, **60**(11), 1226–1240 (2008).

9. Bardhan R., Lal S., Joshi A., and Halas N. J., *Accounts of Chemical Research*, **44**(10), 936–946 (2011).

10. Brazel C. S., *Pharmaceutical Research*, **26**(3), 644–656 (2009).

11. Locher G. L., *American Journal of Roentgenology, Radium Therapy*, **36**, 1–13 (1936).

12. Barth R. F., Vicente M. G., Harling O. K., Kiger W. S., Riley K. J., Binns P. J., Wagner F. M., Suzuki M., Aihara T., Kato I., and Kawabata S., *Radiation Oncology*, **7**, 146 (2012).

13. Hao E., Friso E., Miotto G., Jori G., Soncin M., Fabris C., Sibrian-Vazquez M., M. Graça M., and Vicente H., *Organic and Biomolecular Chemistry*, **6**, 3732–3740 (2008).

14. Gao Z., Walton N. I., Malugin A., Ghandehari H., and Zharov I., *Journal of Materials Chemistry*, **22**, 877–882 (2012).

15. Brozek E. M., and Zharov I., *Chemistry of Materials*, **21**(8), 1451–1456 (2009).

16. Hwang K. C., Lai P. D., Chiang C. S., Wang P. J., and Yuan C. Y., *Biomaterials*, **31**(32), 8419–8425 (2010).

17. Kennedy D. C., Duguay D. R., Tay L. L., Richeson D. S., and Pezacki J. P., *Chemical Communications*, **44**, 6750–6752 (2009).

18. Mortensen M. W., Sorensen P. G., Bjorkdahl O., Jensen M. R., Gundersen H. J. G., and Bjornholm T., *Applied Radiation and Isotopes*, **64**(3), 315–324 (2006).

19. Pickering L., Mitterbauer C., Browning N. D., Kauzlarich S. M., and Power P. P., *Chemical Communications*, **6**, 580–582 (2007).

20. Xu T. T., Zheng J.-G., Wu N., Nicholls A. W., and Roth J. R., *Nano Letters*, **4**(5), 963–968 (2004).

21. Slutsky V. G., Tsyganov S. A., Severin E. S., and Polenov L. A., *Propellants, Explosives, Pyrotechnics*, **30**(4), 303–309 (2005).

22. van Devener B., Perez J. P. L., and Anderson S. L., *Journal of Materials Research*, **24**(11), 3462–3464 (2009).

23. van Devener B., Perez J. P. L., Jankovich J., and Anderson S. L., *Energy Fuels*, **23**, 6111–6120 (2009).

24. Du S. W., Tok A. I. Y., and Boey F. Y. C., *Solid State Phenomena*, **136**, 23–28 (2008).

25. Mortensen M. W., Bjorkdahl O., Sørensen P. G., Hansen T., Jensen M. R., Gundersen H. J. G., and Bjørnholm T., *Bioconjugate Chemistry*, **17**(2), 284–290 (2006).

26. Petersen M. S., Petersen C. C., Agger R., Sutmuller M., Jensen M. R., Soerensen P. G., Mortensen M. W., Hansen T., Bjornholm T., Gundersen H. J., Huiskamp R., and Hokland M., *Anticancer Research*, **28**(2A), 571–576 (2008).

27. Terrones M., Romo-Herrera J. M., Cruz-Silva E., López-Urías F., Muñoz-Sandoval E., Velázquez-Salazar J. J., Terrones H., Bando Y., and Golberg D., *Materials Today*, **10**(5), 30–38 (2007).

28. Chen Y., Chadderton L. T., Gerald J. K., and Williams J. S., *Applied Physics Letters*, **74**(20), 2960–2962 (1999).

29. Chen Y., Conway M., Williams J. S., and Zou J., *Journal of Materials Research*, **17**(8), 1896–1899 (2002).

30. Wang J., Lee C. H., and Yap Y. K., *Nanoscale*, **2**(10), 2028–2034 (2010).

31. Chopra N. G., Luyken R. J., Cherrey K., Crespi V. H., Cohen M. L., Louie S. G., and Zettl A., *Science*, **269**(5226), 966–967 (1995).

32. Ciofani G., Raffa V., Menciassi A., and Dario P., *Journal of Nanoscience and Nanotechnology*, **8**(12), 6223–6231 (2008).

33. Ciofani G., Raffa V., Menciassi A., and Cuschieri A., *Biotechnology and Bioengineering*, **101**(4), 850–858 (2008).

34. Scherer F., Anton M., Schillinger U., Henke J., Bergemann C., Kruger A., Gansbacher B., and Plank C., *Gene Therapy*, **9**, 102–109 (2002).

35. Ciofani G., Raffa V., Yu J., Chen Y., Obata Y., Takeoka S., Menciassi A., and Cuschieri A., *Current Nanoscience*, **5**(1), 33–38 (2009).

36. Ciofani G., Raffa V., Menciassi A., and Cuschieri A., *Nano Today*, **4**(1), 8–10 (2009).

37. Vannucci L., Lai M., Chiuppesi F., Ceccherini-Nelli L., and Pistello M., *New Microbiologica*, **36**(1), 1–22 (2013).

38. Maitra A., *Expert Review of Molecular Diagnostics*, **5**(6), 893–905 (2005).

39. Wang W., Li W., Ma N., and Steinhoff G., *Current Pharmaceutical Biotechnology*, **14**(1), 46–60 (2013).

40. Raffa V., Ciofani G., and Cuschieri A., *Nanotechnology*, **20**(7), 075104 (2009).

41. Huang Y., Lin J., Bando Y., Tang C., Zhi C., Shi Y., Muromachi T., and Golberg D., *Journal of Materials Chemistry*, **20**(5), 1007–1011 (2010).

42. Chen X., Wu P., Rousseas M., Okawa D., Gartner Z., Zettl A., and Bertozzi C. R., *Journal of the American Chemical Society*, **131**(3), 890–891 (2009).

43. Ciofani G., Danti S., D'Alessandro D., Moscato S., and Menciassi A., *Biochemical and Biophysical Research Communications*, **394**(2), 405–411 (2010).

44. Ciofani G., Danti S., Genchi G. G., D'Alessandro D., Pellequer J. L., Odorico M., Mattoli V., and Giorgi M., *International Journal of Nanomedicine*, **7**, 19–24 (2012).

45. Ciofani G., Danti S., D'Alessandro D., Ricotti L., Moscato S., Bertoni G., Falqui A., Berrettini S., Petrini M., Mattoli V., and Menciassi A., *ACS Nano*, **4**(10), 6267–6277 (2010).

46. Ciobanu N., *Biocompatible Magnetic Nano-Clusters Containing Iron Oxide Respectively Iron Oxide–Boron with Primary Use in Magnetic Drug Targeting and Boron Neutron Capture Therapy*, European Patent Application EP2277544 (26.01.2011).

47. Yaghi O. M., O'Keeffe M., Ockwig N. W., Chae H. K., Eddaoudi M., and Kim J., *Nature*, **423**(6941), 705–714 (2003).

48. Efremenko A. V., Ignatova A. A., Borsheva A. A., Grin M. A., Bregadze V. I., Sivaev I. B., Mironov A. F., and Feofanov A. V., *Photochemical and Photobiological Sciences*, **11**(4), 645–652 (2012).

49. Petukhov I. A., Maslov M. A., Morozova N. G., and Serebrennikova G. A., *Russian Chemical Bulletin, International Edition*, **59**(1), 260–268 (2010).

50. Sivaev I. B., and Bregadze V. I., *European Journal of Inorganic Chemistry*, **2009**(11), 1433–1450 (2009).

51. Luguya R., Jensen T. J., Smith K. M., and Vicente M. G. H., *Bioorganic and Medicinal Chemistry*, **14**(17), 5890–5897 (2006).

52. Ol'shevskaya V. A., Nikitina R. G., Savchenko A. N., Malshakova M. V., Vinogradov A. M., Golovina G. V., Belykh D. V., Kutchin A. V., Kaplan M. A., Kalinin V. N., Kuzmin V. A., and Shtil A. A., *Bioorganic and Medicinal Chemistry*, **17**(3), 1297–1306 (2009).

53. Ol'shevskaya V. A., Savchenko A. N., Zaitsev A. V., Kononova E. G., Petrovskii P. V., Ramonova A. A., Tatarskiy V. V., Jr., Uvarov O. V., Moisenovich M. M., Kalinin V. N., and Shtil A. A., *Journal of Organometallic Chemistry*, **694**(11), 1632–1637 (2009).

54. Ratajski M., Osterloh J., and Gabel D., *Anti-Cancer Agents in Medicinal Chemistry*, **6**(2), 159–166 (2006).

55. Bregadze V. I., Sivaev I. B., Lobanova I. A., Titeev R. A., Brittal D. I., Grin M. A., and Mironov A. F., *Applied Radiation and Isotopes*, **67**(7–8 Suppl), S101–S104 (2009).

56. Efremenko A. V., Ignatova A. A., Borsheva A. A., Grin M. A., Bregadze V. I., Sivaev I. B., Mironov A. F., and Feofanov A. V., *Photochemical and Photobiological Sciences*, **11**(4), 645–652 (2012).

57. Matsumura Y., and Maeda H., *Cancer Research*, **46**, 6387–6392 (1986).

58. Maeda H., Sawa T., and Konno T., *Journal of Controlled Release*, **74**(1–3), 47–61 (2001).

59. Sumitani S., and Nagasaki Y., *Polymer Journal*, **44**, 522–530 (2012).

60. Lahiri D., Rouzaud F., Richard T., Keshri A. K., Bakshi S. R., Kos L., and Agarwal A., *Acta Biomaterialia*, **6**(9), 3524–3533 (2010).

61. Chopra N., Gavalas V. G., Bachas L. G., Hinds B. J., and Bachas L. G., *Analytical Letters,* **40**(11), 2067–2096 (2007).

62. Chowdhury R., and Adhikari S., *IEEE Transactions on Nanotechnology*, **10**(4); 659–667 (2011).

63. Ribeiro C., Brogueira P., Lavareda G., Carvalho C. N., Amaral A., Santos L., Morgado J., Scherf U., and Bonifácio V. D. B., *Biosensors and Bioelectronics*, **26**(4), 1662–1665 (2010).

64. Lin Y., Bunker C. E., Fernando K. A. S., and Connell J. W., *ACS Applied Materials and Interfaces*, **4**(2), 1110–1117 (2012).

65. Reddy A. L. M., Gupta B. K., Narayanan T. N., Martí A. A., Ajayan P. M., and Walker G. C., *The Journal of Physical Chemistry C*, **116**(23), 12803–12809 (2012).

66. Pfister A., Zhang G., Zareno J., Horwitz A. F., and Fraser C. L., *ACS Nano*, **2**(6), 1252–1258 (2008).

67. Huang Q., Bando Y., Zhao L., Zhi C. Y., and Golberg D., *Nanotechnology*, **20**(41), 415–501 (2009).

Index

argon atmosphere 87–88
arrangement
  geometrical 15–16
  ordered 5
atomic structure 4, 6–7, 9, 53
atoms
  buckled 42
  bulk boron 105
  hydrogen 17
  metal 37
average electron density 39–40

bacteriochlorins 112
ball-milling method 105
  surfactant-assisted 104
band structures 22, 32, 36, 63–64
  electronic 21, 30, 64
B–B bonds in single-layered sheets 42
BCNT, see boron neutron capture therapy
binding energy 28, 30, 43, 60
biocompatibility 102, 115
Bl boron concentration ratios 103
BNCT 102–3, 105–7, 112–14, 117
BNNTs, see boron nitride nanotubes
BNPs, see boron nanoparticles
BNTs, see boron nanotubes
  mechanical properties of 68
  multiwalled 58, 70–71
  sheet–derived 58
  small-diameter 65
  synthesis 50, 71
bonding manifold 26, 34, 37
bonding states 21, 23, 34
  low-energy 31–32

bonds
  intericosahedral 52, 54
  strong interlayer boron–boron 41
  unique three-center two-electron 78
boron
  concentrated 1
  containing 17, 71, 105, 116
  covalent 2
  dissolved 89
  element 72
  rhombohedral 14, 92
  stabilize hexagonal 24
boron-10 nucleus 103
boron accumulation 114, 117
boron analogues 21
boronated nanoprobes 117
boron atoms/molecules 112
boron bulk crystal 57, 71
boron carbide nanoparticles 104–7
  functionalized 106
  labeled 106–7
boron carbides 105
  available 105
  nanosized 105
boron clusters
  nanoscopic-size 51
  polymerizable 114
boron cluster nanoparticles 113
boron compounds
  encapsulated 114
  tetra-coordinated 2
boron concentrations 103–4
boron-conjugated micelles 115
  obtained 114
boron-containing amino acid 103

boron-containing carborane
moieties 114
boron-containing
nanoheterostructures 102
boron-containing nanoparticles
in medicine 104–5, 107, 109,
111, 113
boron-containing silica 104
boron content, high 104
boron delivery agents 103
potential 104
superior 104
boron doping 24
boron fullerenes 92
stable 26
boron icosahedron 78
boron molecules 24
boron nanobelt 79
boron nanoclusters 3–4, 13–14, 16,
18, 105, 112
neutral 3
quasiplanar 4
boron nanocone 79
boron nanofeathers 87–88
boron nanoparticles (BNPs) 102,
104, 113, 115
boron nanoparticles in biosensing
115
boron nanoparticles in tissue
engineering 115
boron nanoribbons 7
bundled, 7
boron nanosheets 4–5, 92
boron nanostructures in medicine
101–17
boron nanotube samples 9
boron nanotubes (BNTs) 50, 53,
58–59, 68–72, 102
boron nanowire arrays 84, 96
aligned large-scale 95
amorphous 86
boron nanowire patterns 96
boron nanowires
aligned 83, 85

aligned amorphous 85
aligned single-crystal 83
branched 87
field emission properties of 96
growth of 81, 85, 89
growth of amorphous 86, 88
growth of crystalline 10, 82,
89–90
image of 82
patterned 96
properties of 7, 92–93, 95
single 95
single-crystalline 94
boron nanowire samples 93
boron nanowires and nanochains
80
boron neutron capture therapy
(BCNT) 102, 105, 108
boron nitride nanotubes (BNNTs)
107–11, 115, 117
boron nitrides/carbides 102
boron nitride systems 51
boron nucleus, amorphous 86
boron occupies 50
boronophenylalanine 103
boron–organic 112
boron oxide, substoichiometric 86
boron particles 89
boron pellet 90
boron polymorph, enigmatic 13
boron polymorphs 14
boron powder, compacting 88
boron powder concentrations 87
boron-rich nanowires, amorphous
81
boron sheet buckles 39
boron sheet configurations 71
boron sheet lattice 58–59
boron sheet lattice vectors 59
boron sheets
atomic structure 55
buckled triangular 58
buckling 37, 40

distorted hexagonal 55, 57–58, 61–62, 69
few-layered 41
flat triangular 32
hexagonal 21, 57
hexagonal monolayer 6
mixed-phase 30–31, 33, 35
mixed-phase triangular-hexagonal 26–27, 29
monolayer 5–6, 57
multilayered 40
perfect triangular 56
representative double-layered 42
rolled 53
single-layer 17, 54, 71
single-layered 26, 37, 43
single-layer quasi-2D 53
single-layer two-dimensional 53–58
stabilized monolayer 5
stable double-layered 44
structural properties of monolayer 5
thin 21, 23, 25
triangular 57
two-dimensional 19–45, 53, 55
unit–based 55
boron sheet structure 26, 60
boron sheet studies 53
boron structures
    elemental 51, 70
    rhomb-centered hexagonal 10
boron sublattice, stable 37
boron target 90
buckled triangular sheet
    configurations 57
buckling atoms 43
buckling pattern 29–30, 66
bulk, boron crystalline 57, 71
bulk boron 43, 77, 92–93

carbon atoms 21, 68
carbon bulk 57–58, 71

carbon nanostructures 4, 49
carbon nanotubes (CNTs) 49–50, 53, 58–59, 64, 68, 72, 107
carbon nanotubes, single-walled 7, 53
carbon structures, thin 41
carrier
    nonviral 109
    novel boron 115
    unique boron-rich 105
catalytic chemical vapor deposition (CCVD) 107
CCVD, *see* catalytic chemical vapor deposition
cells
    cancer 112–13
    states/unit 8
    unit 6, 26, 36, 42, 59, 66
chemical bonding 13, 15, 17, 34, 54, 57, 69, 71
chemical vapor deposition (CVD) 10, 79, 81–84, 107
chemistry, nanoboron 3
Chinese hampster ovary (CHO) 110
chiral vector 53, 58–59
CHO, *see* Chinese hampster ovary
class, structural 53–54
cluster-based crystal structures 93
clusters
    boron atom 83
    convex 3–4
    larger 20
    planar B6 nanoboron 3
cluster science 13
CNTs, *see* carbon nanotubes
cohesive energy 54, 57, 60, 62, 71
community, scientific 49–50
compound nanotubes 50
compounds, boron–oxygen 93
conductivity, electrical 79, 92–93
configuration, possible 53–54
conjugated boron-containing molecules 112

conjugates, boronated 113
contact, ohmic 92–93, 95
crystalline boron nanowires 10,
    79, 82, 87, 89–90, 92
current data 95
current density 95
    high 94, 96
curvature energies 60–61
CVD, *see* chemical vapor deposition

DA-BNPs 105
delivery, enhanced boron 112
density
    hexagon hole 27, 30, 56
    high boron 104
densities of states (DOSs) 22,
    24–26
density functional theory (DFT) 7,
    9, 21–22, 26, 43, 52, 56, 58–60,
    63–64, 70–71
devices, crystalline boron
    nanowire 92
DFT, *see* density functional theory
DFT calculations 21, 60, 62–64
DFT predictions 63–64
DFT values 62, 69
diseased sites 102, 117
doped boron nanomaterial 112
DOSs, *see* densities of states
double-layered boron sheets 41
    interlayer bonding in 41–43
drug delivery systems 112–13
drugs, boron-based 114

elastic properties 67–68
electric field 95–96
electron-deficient materials 51
electron density 39–40
electron gas 39
electronic bands 64
electronic properties 2, 8, 58, 62,
    65, 67, 92
electronic structure analysis 54
electrons, spin-paired 33–34

elemental boron 13, 17, 21, 50–51,
    53, 57, 71
elemental boron bulk crystalline
    phases 50, 52
elemental boron bulk solids 54
elemental boron solids 51
elements, rare 1
energies
    band 38
    ground-state 60
    relative formation 61
    repulsive 38
    total 21, 38–40, 69–70
energy differences 66–67
energy range 34
EPR effect 113–14

Fe catalyst particle 108–9
Fermi energy 24–25, 38–39
Fermi level for boron 22
field emission 95–96
    stable 94–95
field emission behavior 96
field emission properties 94–95
fluorescence microscopy 107, 109,
    113
folding, zone 62, 64–65
form boron nanowires 86
form interlayer bonds 42–43
form Y-junctions 84

gene delivery 109, 112
generation, singlet oxygen 113
generalized gradient
    approximation (GGA) 21, 64,
    66
gene transfer 109–10
geometries, nanotubular 20
GGA, *see* generalized gradient
    approximation
GGA predictions 66–67
graphene of carbon 54, 57, 62, 64,
    71
graphene sheets 6, 53, 57

curved 19
graphitic carbon structure 107
ground state 2, 20, 28, 37, 41
ground structure, stable 6
growth mechanisms 72, 84, 89,
    92, 96
growth models 96
growth of boron 45, 89
growth of carbon nanotubes 89

hexagonal holes 27, 30, 36–37
hexagonal lattice 22–23, 26, 32, 34,
    51, 57–59
hexagonal sheets, distorted 54, 57
hexagonal structures 22, 24, 27
high-linear-energy-transfer 103

in-plane bonding states 33–34
in-plane states 26, 34
interlayer bonds 43
intravenous injection 114

kinetic energy, electronic 39–40

large-diameter nanotubes 53
laser ablation 10, 79, 88
laser beam 90–91
laser pulses 90
lattice constants 59, 69–70
LDA, *see* local density
    approximation
LEED, *see* low-energy electron
    diffraction
local density approximation (LDA)
    22, 64, 66–67
localization procedures 34–35
low-energy electron diffraction
    (LEED) 5

magnetron sputtering (MS) 79, 85
maximally localized Wannier
    functions (MLWFs) 33–36
mechanical properties, basic 68

methods, postannealing 87
micelles, boronated 114
mixed-phase sheets 27, 29–30,
    32–33
    possible 27
MLWFs, *see* maximally localized
    Wannier functions
models
    atomic 5–6
    schematic 86
modes, radial breathing 9
modulus, elastic 70
molecules
    biological 115–16
    boronated 117
MS, *see* magnetron sputtering
multiwalled BNNTs 108–10, 116

nanochannel alumina (NCA) 83
nanoclusters
    biocompatible iron oxide–
        boron 112
    larger 3–4
nanodevices, futuristic 78–79
nanoparticles
    boronated 104
    inorganic-based 101
nanosheets, oxidation-resistant
    boron nitride 116
nanostructures
    boron-based 102, 115
    boron-containing 107
    complex boron 37
    higher-dimensional boron 4, 6
    investigated boron 10
    quasiplanar boron 4
nanotube diameters, increasing 66
nanowires
    amorphous 87
    growth of 84–85, 89–90
NCA, *see* nanochannel alumina
NCT, *see* neutron capture therapy
neighbors, nearest 23, 27, 31

neutron capture therapy (NCT)
102–3
number, fixed 36, 38

one-dimensional boron
nanostructures 78–79, 95
orbitals
external bonding 15
localized 22, 36–37
orthopedic scaffolds 115
out-of-plane states 23–24, 31, 33

pairs of atoms 23, 31, 35
pairs of identical sheets 43
PDT, *see* photodynamic therapy
PEG, *see* polyethylene glycol
PEI, *see* polyethyleneimine
PEI-BNNTs 108
phases, bulk crystalline 50–51
photodynamic therapy (PDT) 112
platinum-coated sapphire
substrate 89
platinum tip 89
PLL, *see* poly-L-lysine
Poisson ratio 68–69
polyethylene glycol (PEG) 105,
113–14
polyethyleneimine (PEI) 108
poly-L-lysine (PLL) 108, 110
polymeric micelles 101, 112–13
porphyrin derivatives 112–13
porphyrins, boronated 102, 112
precursors, sheets form 6
pristine SWBNT 62
probes
boronated 103
boron-containing 104
process, boron nanowire initiation
84
properties
basic 50, 53, 58–59, 61, 63, 65,
67, 69
magnetic 108
pulse frequency 88–89

quantum, high 113
quasiplanar 3–4

radio-frequency (RF) 85, 105
Raman spectra 8–9
recoiling lithium-7 103
region, central 92–93
reported elemental nanotubes 50
RF, *see* radio-frequency
RF power 87–88
rhombohedral boron crystals 4

SAED pattern 10, 85, 88–89
scanning electron microscope 92,
94
sensitizers, boronated 102
SERS, *see* surface-enhanced Raman
scattering
sheet configuration 57
sheet form 45
sheet plane 37, 66
nominal 24
sheet structures 28
sheet structure, stable boron 56
sheet structures, multiple 53
sheet symmetry 38
silicon, pristine 5
silicon substrate 85, 87, 89
single-layered sheets 41–42
single-walled boron nanotubes
(SWBNTs) 17, 49–50, 52–58,
60, 62, 64, 66, 68, 70, 72
single-walled nanotubes 64
small-diameter nanotubes 64
smallest-diameter nanotube 66
solid boron carbide nanoparticles
105
space, phase 27
squares, green 35–36, 42
stability, energetic 19, 58–59
mixed-phase boron sheets,
stability of 30–31, 33, 35
stabilizing boron bilayers 43

states
    densities of 22, 25
    electronic 6, 35, 56
    localized 35, 37
stiffness, calculated SWBNT 69
strain, axial 69
structural motifs, simple 17, 54, 71
structural robustness 101, 115
structural unit 78
structures
    atomic-scale 19, 45
    buckled 26, 28
    complex 28, 51
    ground-state 3, 30, 37, 72
    metastable 41
    nanotubular 60, 68
    stable 20, 24, 26, 28, 31, 33, 41
    well-defined 17, 54, 71
subunits, structural 5–6
surface, boron-covered 5
surface buckling 63, 65–66, 70
surface-enhanced Raman
    scattering (SERS) 116
surface structures 45, 53
SWBNT-arm 59
SWBNT-chi 59
SWBNT derived 58–59, 61, 63, 65,
    67, 69
SWBNT diameter 62, 69
SWBNTs, *see* single-walled boron
    nanotubes
    armchair 59, 62
    based 61
    buckled-surface 66
    derived 71
    form 66
    large-diameter 64
    sheet–derived 58
    small-diameter 62–63, 66
    small-radii 62, 67, 70
    synthesis of 54, 71
    zigzag 59
SWBNT segment 67
SWBNT-zz 59
SWCNTs 53, 58–59, 62, 68–70

synthesis conditions of boron
    nanowires, 80–81

table, periodic 21, 50, 77–78
target cells 109–10
target tissues 109–10
techniques, zone-folding 62
temperatures, eutectic 84
tetragonal boron bulk 54
theory, elastic 60, 62
therapies, activated 102, 112
thermal neutrons 2, 102–3, 106,
    114
three-center, delocalized two-
    electron 51–52
three-center bonding 31–32
tissues, normal 103
T-junction boron nanowires 87
transport properties, thermal 7
triangular
    buckled 5, 57, 60, 62
    flat 56
triangular lattice 24, 26, 31, 33, 51
triangular regions 26, 36, 66
triangular sheet 24, 26–27, 31–34,
    56
    buckled 54–56
    flat 26, 35, 56
triangular structure 26, 31
tubular structures 8, 60
tumor boron concentration 103
tunable boron content 112

valence electrons 15, 17, 31, 33,
    37, 70
vectors, primary 58

walls, tubular 53
well-aligned single-crystalline
    boron nanowires arrays 83
work function 92–93

Young's modulus 68–70